"A graceful examination of parallels between the United States and the Roman Empire." —*Boston Globe*

"Lucid, learned, witty, and thought-provoking . . . a combination of morality play, cautionary tale, and enlightened plea for the enduring values of civic virtue." —*Chicago Tribune*

"Mordantly funny . . . filled with arresting observations." —*The Nation*

"You painlessly learn a lot about ancient Rome in this smart, briskly paced book, and a lot about contemporary America too." —*Salon.com*

"Uncommonly witty and wise . . . employs Rome's past as a framework to ask trenchant questions about our own present and future."
—*Cleveland Plain Dealer*

"Since the beginning, Americans have nervously compared themselves to Rome. I don't think the exercise has ever been as instructive or entertaining as what Cullen Murphy has produced."
—James Fallows, international correspondent
for the *Atlantic Monthly*

"An urgent, ever-so-timely reminder that what we cherish in America will only be lasting if we make it so."
—Samantha Power, Pulitzer Prize–winning
author of *A Problem from Hell*

"Read in the light of the mess we're in, this is a disturbing book, brimming with hope." —E.J. Dionne Jr., syndicated columnist and
author of *Why Americans Hate Politics*

"Mesmerizing." —*Financial Times*

BOOKS BY CULLEN MURPHY

Rubbish!
The Archaeology of Garbage
(*with William L. Rathje*)

Just Curious: Essays

The Word According to Eve:
Women and the Bible
in Ancient Times and Our Own

Are We Rome?
The Fall of an Empire and
the Fate of America

ARE WE ROME?

The Fall of an Empire and
the Fate of America

Cullen Murphy

A MARINER BOOK
HOUGHTON MIFFLIN COMPANY
BOSTON • NEW YORK

First Mariner Books edition 2008

Copyright © 2007 by Cullen Murphy

www.houghtonmifflinbooks.com

Library of Congress Cataloging-in-Publication Data
Murphy, Cullen.
Are we Rome? : the fall of an empire and the fate of America
/ Cullen Murphy.
p.   cm.
Includes bibliographical references.
ISBN-13: 978-0-618-74222-6
ISBN-10: 0-618-74222-0
1. United States — Civilization — Roman influences.
2. National characteristics, American. 3. United States —
Foreign relations. 4. United States — Territorial expansion.
5. Imperialism. 6. Rome — History. I. Title.
E169.1.M957 2007
970.01 — dc22   2006035717

ISBN 978-0-547-05210-6 (pbk.)

Book design by Melissa Lotfy

PRINTED IN THE UNITED STATES OF AMERICA

VB 10 9 8 7 6 5 4 3 2 1

To my friends and colleagues at
The Atlantic Monthly,
1985–2005

Contents

ARE WE ROME?

Prologue

The Eagle in the Mirror

Urbs antiqua fuit. . . . Urbs antiqua ruit.
There once was an ancient city. . . . The ancient city fell.
— Virgil, *The Aeneid*

IMAGINE THE SCENE: a summer day, late in the third century A.D., somewhere beyond Italy in the provinces of the Roman Empire, perhaps on the way to a city like Sirmium, south of the Danube, in what is now Serbia, where several roads converge — good Roman roads of iron slag and paving stone. The Roman road system is immense — more than 370 separate highways stretching some 53,000 miles all told, about the length of the U.S. interstate system. In these difficult final centuries of the imperium a Roman emperor travels constantly, and his progress makes for quite a spectacle. "The peasants raced to report what they had seen to the villages," a contemporary remembers. "Fires were lit on the altars, incense thrown on, libations poured, victims slain."

The emperor here is perhaps Diocletian, a man of the hinterland, from Dalmatia, and wherever the emperor resides, so resides the imperial government, although Rome itself will long retain its symbolic character — will long be referred to as "the city" even by people five hundred miles away. Who is this Diocletian? No friend of the Christians; he is a "traditional Roman values" man, and his persecutions are intense. But he has restored Rome's stability, at great cost, and in his travels he projects Rome's power. Before the emperor's arrival, advance men known as *mensores* would have been sent ahead to requisition supplies and arrange for security. If you have business with the imperial court, perhaps bearing a petition from your beleaguered city or a plea from your patrician family, and make your way toward the emperor's encampment, you will encounter other supplicants like yourself. Some of them may have been following the emperor for weeks or months. You will also encounter a defensive ring a few miles outside your destination, and find the roads dense with military traffic; and as you draw closer, the character of the armed forces will change, from auxiliaries to legionaries to the imperial bodyguard, a force known as the *protectores*. The imperial eagle flutters on their standards.

At last, in the center, you find the *comitatus* itself, the sprawling apparatus, several thousand strong, that encompasses not only the emperor's household and its personnel — the eunuchs and secretaries, the slaves of every variety (the emperor may own 20,000 of them) — but also the ministries of government, the lawyers, the diplomats, the adjutants, the messengers, the interpreters, the intellectuals. And of course you also find the necessities of life and the luxuries, the rich food and drink. Gone is the simple camp fare of Trajan's day, the bacon, cheese, and vinegar. A letter survives describing the table laid for just one Roman dignitary (and four companions) visiting Egypt — "ten white-head fowl, five domestic geese, fifty fowl; of game-birds, fifty geese, two hundred birds, one hundred pigeons"; multiply accordingly for the emperor and his household. And the ruler himself: How does he spend his time? Receiving petitions? Perhaps he remembers the famous story of one of his predecessors, Hadrian, who put off a pleading woman with the words "I do not have the leisure," only to receive the reply "Then stop being emperor!" (Hadrian made time for the woman.) Consulting with his generals? Repairs to the Danube forts are an urgent necessity, given how many of the German tribes cross over every winter when the river freezes. Dictating letters and decrees? Maybe writing something in his own hand? An earlier emperor, Marcus Aurelius, composed part of his *Meditations* while on a military campaign along the northern frontier; Book One ends with the notation that it was written "among the Quadi," the people he was fighting. Whoever the emperor may be, gathered around the august presence is the imperial government in microcosm, with its endless trunks full of documents; the wagons carrying the treasury and perhaps the mint itself; the blacksmiths and parchment makers; the musicians, courtesans, diviners, and buffoons; the people known as *praegustatores,* who taste the emperor's food before he himself does; the people known as *nomenclatores,* whose job it is to call out the names of the emperor's visitors, and who have given us the word *nomenklatura,* for the core group of bureaucrats and toadies who function within any nimbus of great power. All in all the *comitatus* is, in its way, the cluster of people who in our own time would be encompassed by the Washington e-mail designation "eop.gov."

Or so it occurred to me one summer morning not long ago as my plane touched down in the rain at Shannon Airport, in the Republic of Ireland. The domain name "eop" stands for "Executive Office of the President"—that is, the White House and its extensions—and as it happened, the president of the United States had arrived in Ireland shortly before I did, for an eighteen-hour official visit. His two Air Force One jumbo jets were parked on the shiny tarmac, nose to nose. The presidential eagle, a descendant of Rome's, glared from within the presidential seal, painted prominently near the front door of each fuselage. A defensive perimeter of concertina wire surrounded the two aircraft. Surface-to-air missiles backed it up. The perimeter was manned by American forces in battle fatigues, flown in for the occasion—just one element of the president's U.S. security detail, a thousand strong. Other security personnel peered down from the rooftops of hangars and terminals, automatic weapons at the ready. Ringing the airport was a cordon of Scorpion tanks supplied by the Irish Republic. A traveling president, too, brings with him a government in microcosm. Air Force One can carry much of the presidential *comitatus*—cabinet members and courtiers and cooks, speech doctors and spin doctors. Provisioning has not been overlooked: the plane can serve meals for 2,000 people, the supplies bought anonymously at American supermarkets by undercover agents, the updated version of those *praegustatores*. And if there's a medical emergency? An onboard operating room is stocked with blood of the president's type; his personal physician is at hand. From the plane's command center a president can launch and wage a nuclear war, or any other kind, for that matter. The forward compartment is what passes for a throne room, containing the president's leather armchair and his wraparound oak desk and his telephone with its twenty-eight encrypted lines.

Off in the mist would be the Air Force cargo planes, which had brought helicopters, a dozen Secret Service SUVs, and the official presidential limousine (plus the official decoy limousine), its windows three inches thick and its doors so heavy with armor that gas-powered pistons must be used to help open them. Four U.S. naval vessels plied the Shannon River estuary nearby. Outside the airport the roads were jammed with Irish soldiers and police officers—6,000 in all, slightly more than an entire Roman legion—and on even the tiniest boreens security personnel with communications piglet tails trailing from their ears would emerge from hiding places in the bracken if a passing car, like mine, so much as slowed to avoid some sheep.

Had this president of the United States, George W. Bush, been of a mind to compose his own *Meditations* on this visit, he could legitimately say that he wrote them "among the Alemanni, the Franci, the Celtae," because he was here with the Germans, the French, the Irish, and a number of other tribes for a summit meeting with members of the European Union — a meeting, in other words, with the leaders of allied or subsidiary nations. Ireland, though not technically an American ally, often functions as a client state, and has allowed the United States to route hundreds of transport planes through Shannon Airport, bearing American troops bound for duty in Iraq and Muslim captives bound for interrogation in Eastern Europe and elsewhere. "You are aware of your role as a tributary," a senior British minister has written of his encounters with American officials on occasions like this one (where he was present). "You come as a subordinate bearing goodwill and hoping to depart with a blessing on your endeavors."

The Empire That Won't Go Away

PRESIDENT AND EMPEROR, America and Rome — the comparison is by now so familiar, so natural, that you just can't help yourself: it comes to mind unbidden, in the reflexive way that the behavior of chimps reminds you of the behavior of people. Is it really ourselves we see? Everyone gets it whenever a comparison of Rome and America is drawn — for instance, in offhand references to welfare and televised sports as "bread and circuses," or to illegal immigrants as "barbarian hordes." We all understand what's meant by the thumbs-down sign — *pollice verso,* as the Romans would have said — and know the gladiatorial context from which it came. It's almost compulsory to speak of political pollsters as latter-day versions of Rome's oracles, the *augurs* and *haruspices,* who sought clues to national destiny by studying the flight of birds and the entrails of slaughtered sheep. When a reference is made to an "imperial presidency," or to the president's aides as a "Praetorian Guard," or to the deployment abroad of "American legions," no one quizzically raises an eyebrow and wonders what you could possibly be talking about. To American eyes, Rome is the eagle in the mirror.

Popular culture, the national id, is saturated with references to the Roman Empire. Not long ago HBO and ABC each launched a fictionalized "swords-and-sandals" miniseries set in ancient Rome and centered on the first glimmerings of imperial destiny, as the venerable but creaky Roman Republic began to fall apart. Novels about Rome are reliable bestsellers. The *Star Wars* saga is in essence a Rome-and-America amalgam, about the last remnant of a dying republic holding out against the empire that would supplant it. Earlier films about Rome, such as *The Robe* and *Quo Vadis?*, *Spartacus* and *Ben-Hur*, were crowd-pleasing vehicles that carried implicit political messages against totalitarianism and McCarthyism. (In *The Robe* the emperor Tiberius shows his true colors as an anti-American when he describes the desire for human freedom as "the greatest madness of all.") Liam Neeson, the villain of *Batman Begins*, cites Roman precedent to justify his destruction of Gotham: "The League of Shadows has been a check against human corruption for thousands of years," he tells Bruce Wayne. "We sacked Rome. Loaded trade ships with plague rats."

Rome as a point of reference is not exactly new. Americans have been casting eyes back to ancient Rome since before the Revolution. Today, though, the focus is not mainly on the Roman Republic (as it was two centuries ago, when America was itself emerging as a republic) but as much or more on the empire that took the republic's place. The focus is also as much on the decline and fall of Rome as on its rise and zenith. Depending on who is doing the talking, Rome serves as either a grim cautionary tale or an inspirational call to action. Albert Schweitzer once observed that people setting out to write a life of Jesus all end up looking at their own reflections, as if gazing into the water of a well. In a similar way, those who explore the example of Rome tend to discover that it somehow resonates with their own concerns. I won't pretend to be an exception.

Obviously, the emergence of America as the world's sole superpower, and the troubles it has encountered in that role, explain much of the revival of the Roman Empire in the American imagination. An assortment of "triumphalists" (not their term, of course) see America as at long last assuming its imperial responsibilities, bringing about a global Pax Americana like the Pax Romana of Rome at its most commanding, in the first two centuries A.D. Some form of this idea has been around for decades, and it is here to stay. America's difficulties in Iraq (and in Afghanistan, Iran, Lebanon, North Korea, and elsewhere) are seen as a bump or a challenge—the inevitable price of global leadership—not as a dead end. Charles Krauthammer, the *pontifex maximus* of this outlook, has written: "America is no mere international citizen. It is the dominant power in the world, more dominant than any since Rome. America is in a position to reshape norms, alter expectations, and create new realities. How? By unapologetic and implacable demonstrations of will." William Kristol, the editor of the conservative *Weekly Standard,* ascends to the purple with fewer words: "If people want to say we're an imperial power, fine." The neoconservative writer Max Boot, arguing that America must become the successor empire to Britain (which once saw itself as the successor empire to Rome), has called for "the sort of enlightened foreign administration once provided by self-confident Englishmen in jodhpurs and pith helmets." The triumphalist-in-chief, trading jodhpurs for flight suit, is of course George W. Bush. He has stated that arms races by other nations are "pointless," because American power is now and will forever be kept "beyond challenge" and capable of striking "at a moment's notice in any dark corner of the world."

"Declinists" (also not their term) see this same incipient American empire as dangerously overcommitted abroad and rusted out at home, like Rome in its last two centuries. The historian and columnist Chalmers Johnson, who disparages President Bush as a "boy-emperor," writes in a recent book: "Roman imperial sorrows mounted up over hundreds of years. Ours are likely to arrive with the speed of FedEx." In this view, part of the problem is "imperial overstretch," to use the historian Paul Kennedy's well-known term — our military and globalist ambitions exceed our capacity to pay for them. Another part of the problem is moral and political: empires destroy liberty — always have, always will. Today, the declinists say, the executive branch's imperial need for secrecy, surveillance, and social control, all in the name of national security, is corroding our republican institutions.

Somewhere between the declinists and the triumphalists are those, like the historian Niall Ferguson, in *Colossus,* who argue that at any given moment *some* great power needs to step up and perform the world's various imperial chores — and that the United States is the only one currently available. "Unlike most European critics of the United States," Ferguson writes, "I believe the world needs an effective liberal empire and that the United States is the best candidate for the job." But America, he goes on, is an "empire in denial." It lacks the will and the staying power, the skill and the desire, to shoulder an imperial role. The dispossessed second sons of England's landed gentry and a raft of ambitious and opportunity-starved Scots and Irish lit out for the colonies and there spent their lives governing the British Empire, a sprawling red mass on the maps. America's best and brightest, in contrast, "aspire not to govern Mesopotamia but to manage MTV; not to rule the Hejaz but to run a hedge fund." The problem here, in other words, is "imperial understretch."

The Rome debate has its outright Jeremiahs, its prophets of doom. The social analyst and urban planner Jane Jacobs, in a spirited and hortatory book called *Dark Age Ahead,* published shortly before her death at the age of ninety, all but consigns Western civilization to a new "post-Roman" era of medieval chaos and woe, brought on by the collapse of strong families, the perversion of science, and the oppressive distortion of the government's taxing power. She sees a lethal dynamic at work: "The collapse of one sustaining cultural institution enfeebles others.... With each collapse, still further ruin becomes more likely."

The rot-from-within camp has a conservative flank, too. The classicist and military historian Victor Davis Hanson, sounding like an old Roman, bemoans the American elite's self-absorption, moral relativism, and lack of will. "The anti-Americans often invoke Rome as a warning and as a model, both of our imperialism and of our foreordained collapse," Hanson writes. But, he argues, Rome's situation was more parlous in 220 B.C. (when it faced the challenge of Carthage) than in 400 A.D. (when it faced the barbarians): "The difference over six centuries, the dissimilarity that led to the end, was a result not of imperial overstretch on the outside but something happening within that was not unlike what we ourselves are now witnessing. Earlier Romans knew what it was to be Roman, why it was at least better than the alternative, and why their culture had to be defended. Later in ignorance they forgot what they knew, in pride mocked who they were, and in consequence disappeared."

There are other camps. A group that might be called the "Augustinians," led by the Christian scholar Richard Horsley, wonders if the pursuit of a Pax Americana diverges from the message of Jesus, much as Augustine, in *The City of God*, written shortly after the sack of Rome by Alaric and the Visigoths in 410 A.D., pointed to the incompatibility of earthly and heavenly ambitions. Horsley's views clash with those of "Ambrosians" like the Reverend Jerry Falwell, who see the Pax Americana and the advance of evangelical Christianity as fundamentally inseparable — a throwback to the views on church and state of Ambrose, a Roman prefect and bishop of Milan in the fourth century A.D. "God has raised up America for the cause of world evangelization," Falwell maintains. The idea that an American imperium is part of God's plan was the message of the Christmas card sent out in 2003 by Vice President Dick Cheney and his wife, Lynne.

It read: "And if a sparrow cannot fall to the ground without His notice, is it probable that an empire can rise without His aid?"

And then there are the expansionists, an ironical group who foresee more of the same for America, only bigger and better. In a document that hovers between modest proposal and eccentric manifesto the aging French radical Régis Debray urges the annexation of Europe by America, creating a United States of the West as the only hope against the coming Islamist and Confucianist onslaught. "Who but America can take responsibility, at a reasonable cost, for the peace and unity of the civilized world?" Debray asks. "Do you suppose we would breathe easier under the iron rule of Islam? Or under the domination of China, if by some misfortune she became the only hyperpower?" Referring to an imperial Roman decree of 212 A.D. that extended citizenship to all free men in Rome's provinces, Debray goes on, "I believe the time has come for a new Edict of Caracalla" — meaning American citizenship for Canadians and Mexicans, for Europeans, for Japanese, and for New Zealanders and Australians. (And if it's not too much to ask, can we make sure to include the Caribbean?)

The comparisons, often contradictory, go far beyond military power and global reach. The Roman analogy is cited with respect to the nation's borders and the extent to which America has lost control of them, as Rome lost control of hers. It is cited both by people who see America in the grip of spiritual torpor and sybaritic excess (as Rome at times was) and by those who see it as ruled by moralizing religiosity and outright superstition (as Rome at times also was). It is cited by those who worry about an overweening nationalism and also by those who see an erosion of public spirit.

Cock an ear: you'll hear Rome-and-America analogies everywhere. "It's the fall of Rome, my dear," the food historian Barbara Haber told a reporter when asked about the spread of televised contests featuring gluttony and regurgitation, with their echoes of Roman overindulgence. (Never mind that the fabled *vomitorium* is a myth; the Latin word refers to passageways in amphitheaters that quickly "disgorged" crowds into the streets.) Senator Trent Lott, pushing for the passage of a pork-laden highway bill in 2005, summoned the shades of Rome to his aid. "Part of the reason that Rome eventually collapsed was that it stopped building and maintaining its roads," he argued. "The day we stop investing in better and safer roads is the day we have just one more thing in common with Rome. And Rome fell." In a speech from decades ago that continues to be widely reprinted, Clare Boothe Luce railed against America's anything-goes "new morality" toward sex, conjuring the forlorn attempts of Augustus, Rome's first emperor, to bolster the Roman family in the face of similar licentiousness. ("It was already too late," Luce concluded darkly.) Most people are aware that the Roman Empire was eventually split into western and eastern halves, the one Latin-speaking and centered on Rome, the other Greek-speaking and centered on Constantinople. It's probably only a matter of time before someone sees in this a foreshadowing of the emergence of Red and Blue America.

The larger question still hangs in the air: Are we Rome? That question leads to others: Does the fate of Rome tell us anything useful about America's present or America's future? Must decline and fall lurk somewhere ahead? Can we learn from Rome's mistakes? Take heart from Rome's achievements? And by the way, what exactly *was* the fate of the Roman Empire? Why do historians lock horns over the question, Did Rome really fall?

If you're looking for reasons to brush comparisons aside, it's easy enough to find them. The two entities, Rome and America, are dissimilar in countless ways. It's hard even to know what specific moments to compare: the American experiment is in its third century, and the Roman state in the West spanned more than a millennium, from the eighth century B.C. to the fifth century A.D. Over time, Rome and America molted more than once from their previous selves. But I'll argue that some comparisons do hold up, though maybe not the ones that have been most in the public eye. Think less about decadence, less about military might, and more about how our two societies view the outside world, more about the slow decay of homegrown institutions. Think less about threats from unwelcome barbarians, and more about the healthy functioning of a multi-ethnic society. Think less about the ability of a superpower to influence everything on earth, and more about how everything on earth affects a superpower.

I'll argue further that the debate over Rome's ultimate fate holds a key to thinking about our own. The status quo can't be flash-frozen. A millennium hence America will be hard to recognize. It may not exist as a nation-state in the form it does now — or even exist at all. Will the transitions ahead be gradual and peaceful or abrupt and catastrophic? Will our descendants be living productive lives in a society better than the one we inhabit now? Whatever happens, will valuable aspects of America's legacy weave through the fabric of civilizations to come? Will historians someday have reason to ask, Did America really fall?

Comparative Anatomy

FIRST, LET'S EASE one issue to the side. There are exceptions, but most historians who teach in colleges and universities are skeptical of trying to draw explicit "lessons" from history. No historical episode is precisely like any other, they point out, so no parallel can ever be exact. Too often, they say, people focus on a handful of similarities and ignore all the differences. Worst of all, history gets hijacked for ideological reasons, as when American officials cite the appeasement at Munich to get our armies marching, or the quagmire of Vietnam to keep our armies home. Even when people try to learn sensibly from the past, they may be deriving conclusions that have no relevant application: that's what the charge "fighting the last war" is all about. In their book *Thinking in Time* the historians Richard Neustadt and Ernest May offer a dozen case studies, drawn from foreign affairs, of how history was inadvertently or cynically misapplied by American leaders—if historical thinking was engaged in at all. (They also wonder how Lyndon Johnson, struggling with Vietnam, would have reacted if his national security advisor had ever invoked the example of the Peloponnesian War; and they point out that the monarchs and ministers who led Europe into the carnage of World War I knew the Greek lessons through and through.) Given all this, many historians conclude, scholars have their hands full just trying to figure out what actually happened way back when, a task that in itself may be beyond our meager powers. The British historian A.J.P. Taylor used to say, "The only lesson of history is that there *are* no lessons of history." You can almost visualize his regal verdict: *Pollice verso* —thumbs down!

Of course, many people can't believe their ears when they hear historians talk this way. Not long ago at the University of Chicago a panel of classicists held forth on why it's a mistake to palpate the past for guidance in the present. They ran into opposition from an uncomprehending audience of well-educated non-academics, whose reaction can be summed up as "But, but . . . but how can you say that?" The public has long been schooled to think that being aware of history—and taking historical analogies into account—is actually the smart thing to do. The famous Santayana maxim about what happens to those who forget history is drilled into you by the sixth grade, and everyone who learns it is condemned to repeat it. The Pentagon, taking this idea to heart in a literal-minded and almost endearing way, runs a Center for Army Lessons Learned, at Fort Leavenworth. It maintains a database called the Joint Universal Lessons Learned System. And then there's the example of our own lives, the retort of Everyman: What's the point of "experience," that much-recommended quality, if you can't, or shouldn't, learn something from it?

The scholars are right to be wary; in many ways the history of History is a saga of its misuse. At the same time, as some warn, to rule out any hope of lessons risks making history, especially classical history, into little more than a theme park. The commonsense approach is the one suggested by Carl Becker in a famous lecture to his fellow historians many decades ago: "Mr. Everyman is stronger than we are, and sooner or later we must adapt our knowledge to his necessities. Otherwise he will leave us to our own devices, leave us it may be to cultivate a species of dry professional arrogance growing out of the thin soil of antiquarian research." Becker went on to caution that the whole historical enterprise is treacherous territory indeed: the past plays tricks on the present, and vice versa. But he wasn't ready to throw in the towel.

So again: Are we Rome? One way to answer the question is by assembling a crude ledger of comparisons. My own would start as follows: Leaving aside the knotty and partly semantic issue of what an empire is, and whether the United States truly is one, Rome and America are the most powerful actors in their worlds, by many orders of magnitude. Their power includes both military might and the "soft power" of language, culture, commerce, technology, and ideas. (Tacitus said of the seductive amenities brought to Britain by Rome, "The simple natives gave the name of 'culture' to this factor of their slavery.") Rome and America are comparable in physical size — the Roman Empire and its Mediterranean lake would fit inside the three million square miles of the Lower Forty-eight states, though without a lot to spare. Both Rome and America created global structures — administrative, economic, military, cultural — that the rest of the world and their own citizens came to take for granted, as gravity and photosynthesis are taken for granted. Both are societies made up of many peoples — open to newcomers, willing to absorb the genes and lifestyles and gods of everyone else, and to grant citizenship to incoming tribes from all corners of the earth. And because of this, the identities of both change organically over time. Romans and Americans revel in engineering prowess and grandiosity. Whenever I see the space shuttle, standing upright and inching slowly on its crawler toward the launching pad, I think back to the Rome of Hadrian's day, and the gargantuan statue of the Sun-God, as tall as the shuttle, being dragged into place by twenty-four elephants.

Romans and Americans can't get enough of laws and lawyers and lawsuits. They believe deeply in private property. They relish the ritual humiliation of public figures: Americans through comedy and satire, talk radio and Court TV; the Romans through vicious satire, to be sure, but also, during the republic, by means of the *censoria nota,* the public airing, name by name, of everything the great men of the time should be ashamed of. Romans and Americans accept enormous disparities of wealth, and allow the gap to widen. Ramsay MacMullen, one of the most prominent modern historians of Rome, has said that five centuries of imperial social evolution can be reduced to three words: "Fewer have more." Both Romans and Americans treat the nouveaux riches with lacerating scorn, perhaps concealing hints of admiration. (Think of the character Trimalchio in the *Satyricon* of Petronius; and remember that Fitzgerald's original title for *The Great Gatsby* was *Trimalchio in West Egg.*) Both see themselves as a chosen people, and both see their national character as exceptional. Both recover from colossal setbacks, and both endure periods of catastrophic leadership (though when it comes to murderous insanity, some of Rome's emperors set the bar very high). Both Rome and America look back to an imagined nobler, simpler past, and both see the future in terms of Manifest Destiny. The Romans spoke of having been granted *imperium sine fine* — an empire without end. The American dollar bill uses Rome's own language, and words derived from Virgil, to proclaim a *novus ordo saeclorum* — a new order of the ages. When the first President Bush, after communism's fall, proclaimed the advent of a "new world order," his new rhetoric was actually very old.

But Rome in all its long history never left the Iron Age, whereas America in its short history has already leapt through the Industrial Age to the Information Age and the Biotech Age. Wealthy as it was, Rome lived close to the edge; many regions were one dry spell away from famine. America enjoys an economy of abundance, even surfeit; it must beware the diseases of overindulgence. Rome was always a slaveholding polity, with the profound moral and social retardation that this implies; America started out as a slaveholding polity and decisively cast slavery aside. Rome emerged out of a city-state and took centuries to fully let go of a city-state's methods of governance; America from very early on began to administer itself as a continental power. Rome had no middle class as we understand the term, whereas for America the middle class is the core social fact — our ballast, our gyroscope, our compass. Rome had a powerful but tiny aristocracy and entrenched ideas about the social pecking order; even at its most democratic, Rome was not remotely as democratic as America at its least democratic, under a British monarch. In Roman eyes the best way to acquire wealth was the old-fashioned way, by inheriting it; the Romans looked down on entrepreneurship, which Americans hold in the highest esteem, and despised manual labor. Rome desired foreign colonies and protectorates and moved aggressively to acquire them; America with few exceptions prefers to extend its reach by other means. Rome was economically static; America is economically transformative. For all its engineering skills, Rome generated few original ideas in science or technology; America is a hothouse of innovation and creativity. Despite its deficiencies, as we may perceive them, Rome flourished as a durable culture for more than a thousand years, and acted as a great power for six centuries; whether America has that kind of staying power remains to be seen.

As individuals, Romans were proud, arrogant, principled, cruel, and vulgar; Americans are idealistic, friendly, heedless, aggressive, and sentimental (but, yes, often vulgar, too). I'm not sure that Americans, cast suddenly back in time, would ever warm to second-century Rome, the way they might to Samuel Johnson's London. In their mental maps, their intellectual orientations, their default values, Romans and Americans are further apart than most people suspect. Romans were as bawdy as Americans are repressed. Roman notions of personal honor and disgrace, and the behavior appropriate to each, have no real counterpart in America; Roman officials would unhesitatingly commit suicide in situations that wouldn't make Americans even sit down with Barbara Walters (much less consider resigning). On basic matters such as gender roles and the equality of all people, Romans and Americans would behold one another with disbelief and distaste. The fully furnished frame of mind of a modern American differs hugely from that of a colonial American at the time of Bunker Hill, and even more from that of a settler in Jamestown; the distance between the modern American mind and the ancient Roman one is hard to bridge. If the past is another country, then Rome is another planet. And yet, that planet colonized the one we inhabit now.

Six Parallels

ROME WILL ALWAYS speak across the centuries, and it is too large a thing not to be heard. Like the Bible or the works of Shakespeare, the history of Rome encompasses the whole of the human condition: every motivation, every behavior, every virtue, every vice, every outcome, every moral. And like the Bible and Shakespeare, what Rome has to say is shaped by the listeners in any age.

What is Rome saying to us today? In the pages ahead I'll focus on a half dozen issues for which the example of Rome provides parallels of direct relevance for America. This isn't meant to be a capsule history of the Roman Empire; any number of important subjects, such as religious belief, economic policy, and palace politics, come up only in passing. And it's not meant to highlight every point of contrast between Rome and America; the emphasis is on comparisons that compel attention because there's something to them. Some of the parallels have to do with how Rome and America function on the inside; others have to do with outside pressures and constraints. The parallels aren't fixed in place, and they don't point to an inescapable future. Taken as they are, though, they trace a path that leads to foreseeable consequences — a path, after all, that Rome has already been down.

One parallel involves the way Americans see America; and, more to the point, the way the tiny, elite subset of Americans who live in the nation's capital see America — and see Washington itself. Rome prized its status as the city around which the world revolved. Official Washington shares that Ptolemaic outlook. Unfortunately, it's not a self-fulfilling prophecy — just a faulty premise. And it leads to an exaggerated sense of Washington's weight in the world: an exaggerated sense of its importance in the eyes of others, and of its ability to act alone. Washington led the fight against some of the twentieth century's most dangerous "-isms." Solipsism is one it missed.

Another parallel concerns military power. This is the subject that comes most often to mind when Rome and America are compared. All that empire talk! Rome and America aren't carbon copies or fraternal twins, in either their approach to power or the tools at their disposal. Amid all the differences, though, two large common problems stand out. One is cultural and social: the widening divide between military society and civilian society. The other is demographic: the shortage of manpower. For a variety of reasons, Rome and America both start to run short of the people they need to sustain their militaries, and both have to find new recruits wherever they can. Rome turned to the barbarians for help: not a good long-run solution, history would suggest. America is increasingly turning to its own outside sources — not the Visigothi and the Ostrogothae but the Halliburtoni and the Wackenhuti. Also not a good long-run solution.

A third parallel is something that can be lumped under the term "privatization," which can often also mean "corruption." Rome had trouble maintaining a distinction between public and private responsibilities — and between public and private resources. The line between these is never fixed, anywhere. But when it becomes too hazy, or fades altogether, central government becomes impossible to steer. It took a long time to happen, but the fraying connection between imperial will and concrete action is a big part of What Went Wrong in ancient Rome. America has in recent years embarked on a privatization binge like no other in its history, putting into private hands all manner of activities once thought to be public tasks: collecting the nation's taxes, patrolling its streets, defending its borders. This may make sense in the short term — and sometimes, like Rome, we may have no choice in the matter. But how will the consequences play out over decades, or centuries? Badly, I believe.

A fourth parallel has to do with the way Americans view the outside world — the flip side of their self-centeredness. Rome often disparaged the people beyond its frontiers, and generally underestimated their capabilities, even as it held an outsize opinion of its own superiority and power. America's attitude is more complicated than Rome's, and often more idealistic and well-meaning, but in many ways it's strikingly similar, and it leads to the same preventable form of blindness: either we don't see what's coming at us, or we don't see what we're hurtling toward.

And then, fifth, there is the question of borders. Historians in recent decades have invested much effort in the study of Rome's frontiers, showing that the fringe of empire was less a fence and more a threshold — not so much a firm line fortified with "Keep Out" signs as a permeable zone of continual interaction, sometimes troublesome but normally peaceful and mutually advantageous. The borderlands could hardly have been anything else: this is always the dynamic when a rich and powerful civilization bumps up against a poor and less developed one. The dynamic can't be argued with or neutralized, and yet Rome coped successfully with this reality for many centuries, assimilating newcomers by the millions: that's the happy lesson. When historians describe life along the Rhine or the Danube frontier in Roman times, an American reader can't help conjuring an image of another boundary zone: the one that includes the Rio Grande.

Finally, sixth, comes the complexity parallel. Sprawling powers like Rome and America face a built-in problem. They inevitably become impossible to manage, because the very act of managing has unpredictable ripple effects, of global scale, which in turn become part of the environment that needs to be managed. The theologian Reinhold Niebuhr was writing about a newly predominant America, but his observation (made fifty years ago) applies equally to Rome: "The same strength which has extended our power beyond a continent has also . . . brought us into a vast web of history in which other wills, running in oblique or contrasting directions to our own, inevitably hinder or contradict what we most fervently desire." The bigger the entity and the more things it touches, the more susceptible it is to forces beyond its control. Maintaining stability requires far more work than fomenting instability. Analysts of modern terrorism wring their hands over a version of the same dilemma: governments can win only by defending everywhere; terrorists can win by succeeding anywhere. The complexity problem may have no real solution other than Thoreau's deceptively easy one: "simplify."

The example of Rome instills one more thing — not so much a lesson as a state of mind. It encourages an appreciation for the workings of time itself: patient, implacable, and very, very long. This state of mind can induce a form of resignation. (Isabel Archer, in *The Portrait of a Lady,* is drawn to Rome for just that reason: ("In a world of ruins the ruin of her happiness seemed a less unnatural catastrophe.") Or it can put you on high alert. Time achieves revolutions by invisible increments. Changes that seem inconsequential over a single lifetime can upend the social order over three or four. We don't naturally think in these terms; we're all hemmed in by our one-lifetime horizon. But Rome has a way of raising the vantage point, altering your perspective.

Fly Through

I FIRST VISITED ROME at the age of twelve, some forty years ago, and have been back a dozen times. It is strange to think that the emperor Diocletian himself didn't see Rome until he was in his mid-sixties, older than I am now — he was born elsewhere, always away on business, and didn't visit the greatest city in the empire until celebrating the twentieth year of his reign. (And he didn't like it, found it extravagant, crude, out of control: Las Vegas meets *Blade Runner.*) Nowadays it is possible to visit ancient Rome remotely, on computerized "fly-through" tours that allow you to zoom in to a three-dimensional reconstruction of the ancient Forum. The aim, eventually, is to render the city in full, layer by layer, archaeologically correct to the greatest imaginable degree. Here you are in, say, the third century A.D., on the Campus Martius, the Field of Mars, near the Mausoleum of Augustus, not far from the marble Ara Pacis, the Altar of Peace, the great monument to Augustan tranquillity. And now, a little to the east, here you are in the Forum of Trajan, above it the largest marketplace in the world — the Mall of Romanita, we'd call it today. And after that, outside the monumental city center, you wander among the six-story tenement buildings known as *insulae,* or "islands," teeming and squalid and prone to fire, the ground floors everywhere crowded with shops. Press a key and the years peel away: you can see the city at the time of Marcus Aurelius, in the second century A.D., or, earlier, of Hadrian, or

Augustus, or Julius Caesar, or Cato the Elder, or at the time of the kings, before the sixth century B.C. Watch the walls contract, the temples disappear, the imperial McMansions on the Palatine shrink into republican villas, and the villas into huts, until all that is left on the wooded hill is the legendary cave where Romulus and Remus, the founders of Rome, were suckled by a she-wolf.

I remember the last day of my first trip to Rome, a trip undertaken in real time and space; I walked alone in the early morning with a sketchbook. The ancient paving stones of the Via Sacra were polished with wear, the sun already promising heat, the postcard sellers setting up their stalls like the guides and hawkers who catered to tourists in the very same spots two thousand years before. This was in the mid-1960s, before the thousands of cats that inhabited Rome's ancient precincts had been removed, and the landscape was subtly animated by slurry pixels of feline movement.

I made a drawing that morning, which I still have, of the three standing columns of the Temple of Castor, above the reflecting pool. I remember thinking as I worked, looking up occasionally at the ruined hillside of the Palatine in the near beyond, about the layer cake of happenstance connecting Then and Now. Years later, I came across the fantasy version of that schoolboy reverie: the comparison attempted by Sigmund Freud between the human psyche and the archaeology of Rome. "Historians tell us," he begins, "that the oldest Rome was the *Roma Quadrata*, a fenced settlement on the Palatine. Then followed the phase of the Septimontium, a federation of the settlements on the different hills; after that came the city bounded by the Servian Wall; and later still, after all the transformations during the periods of the republic and the early Caesars, the city which the emperor Aurelian surrounded with his walls." Now imagine, he goes on, that rather than each stage being obliterated by the next, they all somehow survive. Thus it would be possible to see not only the Colosseum in all its grandeur but also the lake in front of Nero's palace, which it replaced. We could gaze simultaneously at today's Castel Sant'Angelo, an imposing umber fortress, and at the bright marble tomb of Hadrian, crowned with a grove of trees, from which the fortress grew. Could our minds be something like that — a psychic device "in which nothing that has come into existence will have passed away, and all the earlier phases of development continue to exist alongside the latest one"?

Well, no, that magic doesn't really work for the mechanics of the mind, Freud decided. But it does work for the way we accrue perceptions of history: as an exercise, we can set anything alongside anything else. If Rome and America can exist simultaneously, why not try to look at them that way?

1

—⁓—

THE CAPITALS
Where Republic Meets Empire

Remember, Roman, that it is yours to lead
other people. It is your special gift.
— Virgil, *The Aeneid*

We are the indispensable nation. We stand tall.
We see further into the future.
— former secretary of state Madeleine Albright

THE EMPIRE of the Romans in the West, its origins trac-
ing back more than a thousand years, drew its last breath
in 476 A.D., when a barbarian army led by a warrior named Odoacer, half Hun and half Scirian, defeated an imperial army that
his barbarians had only a few months earlier been a part of. Odoacer captured and killed the imperial commander. He entered
the city of Ravenna, then serving as an imperial capital, and deposed a youngster named Romulus Augustus, who had reigned
as emperor for little more than a year. Odoacer was scarcely less
worthy of authority than many previous usurpers. He was in
fact well schooled in the ways of Rome, and he was a Christian,
as most Romans by then were. There was no social implosion
after he seized power, no rape and pillage. Rome didn't "fall"
the way Carthage had, six centuries earlier, when the Romans
slaughtered the inhabitants and razed the city, or the way Ber-

lin would, fifteen centuries later, blasted into rubble. Rome itself wasn't touched on this occasion, and throughout the former
empire life went on, little different for most people in 477 from
what it had been in 475. Many regions had been autonomous for
years, under barbarian rulers who gave lip service to the titular
emperor. In Italy the Roman bureaucracy continued to sputter
along.

What changed was this: Odoacer was not recognized as legit-
imate by the eastern emperor, in Constantinople. There would
never be another emperor of the West. The historical symme-
try is almost too good to be true — that the last emperor's name,
Romulus, should also be that of Rome's founder. (Imagine if
the demise of America were to occur under a president named
George.) But more than symbolism was at play. Odoacer understood full well that something had come to an end: he declared
himself king of Italy, and sent the imperial regalia of the Western empire to Constantinople. The pretense of Western unity
was abandoned. Europe would now become a continent of barbarian kingdoms — in embryo, the Europe of nation-states that
exists today.

Thirteen centuries later, on a gloomy evening in 1764, gazing
out from a perch on the Capitoline Hill, above the overgrown
debris of central Rome, Edward Gibbon was seized with a sense
of loss as he contemplated the collapse of a civilization. Monks
sang vespers in a church nearby. Gibbon resolved at that moment
to undertake the great project he would call *The Decline and Fall
of the Roman Empire.* "I can neither forget nor express the strong
emotions which agitated my mind as I first trod, with lofty step,
the ruins of the Forum," he later wrote. A decade after this twi-
light epiphany Gibbon's restless pen evoked the collapse of the
empire: "Odoacer was the first barbarian who reigned in Italy,
over a people who had once asserted their just superiority above
the rest of mankind. . . . The least unfortunate were those who
submitted without a murmur to the power which it was impos-

sible to resist." Gibbon's life was in many ways a sad and lonely one, but *The Decline and Fall of the Roman Empire* was recognized at once as a masterwork, its sonorous cadences enlivened with a dry and biting wit. He observes gratuitously of a monk named Antiochus, for instance, that "one hundred and twenty-nine homilies are still extant, if what no one reads may be said to be extant." Although his picture of the fall may be more cataclysmic than the immediate reality seems to have been, Gibbon established for people ever after that a page of history had been decisively turned. In the West, "decline and fall" has been a catchphrase and a source of anxiety ever since.

The city of Washington, of course, also has a Capitoline Hill — Capitol Hill, named explicitly for its Roman forebear. The view to the west takes in a vast expanse of classical porticoes and marble monuments; gilded chariots and curtained litters would not seem out of place against this backdrop. Washington rose out of a malarial marsh on a river upstream from the coast, as Rome did. Its people, like the Romans, flee the sweltering city in August. The Romans cherished their myth of origin, the story of Romulus and Remus, and on the Palatine Hill you could be shown a thatched hut said to be the hut of Romulus — yes, the very one. Washington doesn't have anything quite like the hut of Romulus, but on Capitol Hill you can find sacred national touchstones of other kinds, such as the contents of Lincoln's pockets when he was assassinated. (They're in the Library of Congress.) Washington resembles Rome in many ways. The physical similarities are visible to anyone. The similarities of spirit are more salient. Materialistic cultures easily forget that "mental outlook" is not some limp and passive construct, of interest chiefly to anthropologists. Mental outlook can drive events and change the world, as the rise of militant Islam makes plain. Washington, too, has been animated by a special outlook. Long ago it was a notion of republican virtue that Romans of an early era would immediately have recognized. Today it's a strutting sense of self and mission that Romans of a later era would have recognized just as readily. Foreigners are well aware of this outlook, friends and enemies alike. It's a pungent quality — an internal characteristic that gives rise to outside counterforces.

The comparison with Rome has always been on the minds of leaders in America's capital. It was celebrated when Washington was no more than a street plan, and inspired what might be called the Bad Virgil school of patriotic verse. ("On broad Potowmac's bank then spring to birth, / Thou seat of empire and delight of earth!") In the settlement's early years there was a tributary of the Potomac called Goose Creek; its name was changed by an aspirational local planter to Tiber Creek. The Jefferson Memorial, off on the Potomac River's edge, is a diminutive version of the Pantheon. Union Station, just below the Capitol, was inspired by the Baths of Diocletian. The Washington Monument recalls the obelisks brought to Rome after the conquest of Egypt. Colonnaded government buildings stretch for miles.

I doubt I'm the only person who has trod, with lofty step, the sculpted gardens of the Capitol and been seized with a vision of how the city below might appear as a ruin. The Washington Monument — imagine it a millennium hence, a chipped and mottled spire, trussed with rusting braces. The stern pile of the Archives building, the Tomb of the Unknown Soldier at Arlington, the gothic National Cathedral on its distant hilltop, the turreted Smithsonian Castle on the Mall — they somehow invite you to see them as derelicts, rendered into darkly impish engravings by the hand of some future Piranesi. What calamity could bring the capital to this condition? Earthquake? Pestilence? Pride? The end of air conditioning?

What Went Wrong

A PAGE OF HISTORY may have been turned in 476 A.D., at least for literary purposes, but it's not easy to pinpoint the moment of Rome's fall—just as, one day, it may not be easy to pinpoint the moment of America's. In many ways, "Rome" had already fallen—had evolved into something different from what it once was, and not always through violence—well before it ceased to exist as a formal political entity. It had once been pagan and by the end was largely Christian. A proud army made up of Romans had long since turned into a paid army made up of barbarians. A republic sustained by flinty yeomen had become a precarious autocracy administered by grasping bureaucrats. At the same time, in very concrete ways Rome didn't fall for centuries, if at all. The eastern half of the empire, the richest and most populous part, centered on Constantinople, survived for nearly another thousand years. In the realms of culture and law and infrastructure and language, the Roman Empire has endured much longer. We still use its alphabet, exploit its literary genres, inhabit its cities, preserve its architectural styles, and follow its schedule of holidays. In many respects the Catholic Church survives as a graft on the empire's stump. *"Non omnis moriar"* ("I shall not wholly die"), the Roman poet Horace proclaimed in one of his most famous odes. He was referring to his work, but he could just as well have been referring to the legacy of his civilization.

Rome began as a farming settlement on hilly portions of the eastern bank of the Tiber River. Tradition puts its founding at 753 B.C., and the Romans calculated the passage of years *ab urbe condita*—"from the founding of the city." The legendary origins of the Roman people go back even further, to the Trojan hero Aeneas, who with family and friends made his way to Italy after the fall of Troy. It's not easy to infer, even when tramping the most ancient parts of today's Rome, what the early settlement was like. Some musty nineteenth-century guidebooks are actually good at reconstructing the landscape—how steep and rugged the seven hills were, and how willows grew here and oak trees there, and where the ferries crossed the marshes, and how high the Tiber floods could rise. Several centuries as a monarchy gave way, in the sixth century B.C., to a republic, with a senate, consuls, and popular voting for some offices. The territory of Rome gradually expanded to encompass the rest of the Italian peninsula and outposts along the Mediterranean coast. With the conclusion of the Third Punic War against its great rival, Carthage, in northern Africa, in 146 B.C., Rome effectively controlled the Mediterranean world. It continued to grow in all directions, impelled by its military prowess, its administrative genius, and its compulsive sense of destiny.

The republic came to a de facto end in 31 B.C., after a century of social turmoil, constitutional crisis, and civil war. Vast Roman armies had thrown themselves against one another across the Mediterranean world. Emerging supreme from the carnage was Octavian, Julius Caesar's grandnephew, who in proto-Orwellian fashion symbolically "restored" the republic while in fact inaugurating the principate, a regime of one-person rule. The outward forms of republican government would be preserved in various ways right to the very end, a progressively meaningless nod to the past, but whatever the disavowals, Rome was now an imperial state. By the second century A.D., the all-powerful emperor ruled a domain that stretched from Scotland to the Sahara, from the Atlantic to the Euphrates. Durable roads linked major and minor cities, defining routes still used today. Seaborne traffic flourished. In the absence of reliable records, one indirect way to gauge the growth in maritime commerce is through underwater archaeology, measuring the change over time in the number of Mediterranean shipwrecks — analogous to tracking the advance of industrialization by the level of pollutants in arctic ice cores. A survey of Roman-era wrecks off the coasts of Italy, France, and Spain yields only about fifty from the period 400–200 B.C., but three times that many from the first two centuries A.D. Ancient Rome held more than a million people; no European city would come close to it in size until the London of Shakespeare's time. The walls that eventually surrounded Rome extend nearly thirteen miles. I once spent the better part of two days walking the entire circuit, noting where the nineteen roads had entered the city, and the eleven aqueducts, and thinking how formidable those walls, forty feet high, must have looked to Alaric and his Visigoths. Little wonder that Alaric cut a deal to get inside.

The decline of Rome came in many forms — in military power, in civil order, in eloquence, in philosophy, in architecture, in trade. Going back to those shipwrecks: they fall off sharply after 200 A.D., and after 400 drop to the levels of half a millennium earlier. It would be a thousand years before seaborne trade returned to the Augustan level. Infrastructure started to degrade: there came a point when full-length columns of colored marble, which literally held up the empire, could no longer be transported to Rome from Greece and Turkey and Egypt. Agricultural methods deteriorated: archaeology indicates that cattle were smaller in the early Middle Ages than they were at the empire's prime. Although 476 has been accepted since Byzantine times as the moment of Rome's demise, there's an element of parlor game in the discussion. Maybe the end *really* came in 455, when the Vandals sacked Rome. Or maybe it came in 410, when the Visigoths sacked Rome. Or in 378, when a great Roman army was destroyed by barbarians at Adrianople. A racist theoretician in Nazi Germany discerned Rome's "first step toward chaos" in a law passed in the fifth century B.C. permitting patricians, the highest social class, to marry plebeians, the lowest.

"Let students of Rome's decline imagine themselves as medical examiners who have been confronted with a corpse," writes the classical historian Donald Kagan. "It is their duty first to establish the time of death and then the cause. It soon becomes apparent that the various historical practitioners who have examined the Roman remains have achieved remarkably little agreement on either question." In 1980, a German historian set out to catalogue all the explanations for the fall of Rome ever pro-

posed, which include degeneracy and deforestation, too much
bureaucracy and too much Christianity. (He cited 210 theories in
all.) The Romans themselves continually lamented the unhappy
state of their society — as Americans compulsively do — even
under circumstances that in retrospect were not all that bad.
"Now we suffer the evils of a long peace. Luxury hatches terrors
worse than wars." That's Juvenal, writing in the second century
A.D., when the collapse of the Western empire lay more than
three centuries ahead. Other writers were bizarrely sanguine, although trouble was just around the corner. "There will never be
an end to the power of Rome," wrote the court poet Claudian,
shortly before the city's sack by the Visigoths. Part of the problem of explaining "decline" is that, like "rise," it doesn't happen
everywhere at the same rate or in the same way. Ronald Reagan
declared the 1980s to be "morning again in America," but dawn
looked a lot different in Silicon Valley than it did in Youngstown.

Still, looking at the range of explanations provides a montage
of Rome's condition. There is, to begin with, the growing number of incursions into the empire by non-Roman peoples — that
is, by the barbarians. Rome had always been adept at assimilating
newcomers; until the rise of America, it was history's most suc-
cessful multi-ethnic state. But the influx eventually became too
much to handle, as the Huns, sweeping out from central Asia,
drove more and more people south and west in front of them,
and finally across the Rhine and the Danube and into the empire. Another explanation: perhaps the culprit was simply a hol-
lowed-out military, whose capacities were no longer up to the
challenge of keeping the barbarians at bay. Related to this: Did
a creeping pacifism come into play? ("We Christians defend the
empire by praying for it," wrote one early theologian.) Some historians blame economic stagnation for the fall of Rome, or corruption, or manpower shortages, or the exhaustion of the soil,
or the depredations of plague, or the more generalized problem,
in one view, that "too few producers supported too many idle

mouths." An implicitly eugenic argument points to the deple-
tion of the elites by centuries of war and civil strife. You could
look at the debilitating effects of a decline of civic spirit. Or at
the rise in taxes, which took a greater toll on ordinary people
than it did on the rich and influential, worsening an already in-
vidious class divide. There was the impact of slavery, whose
harmful consequences were moral and psychological as well as
economic. And there was the chaos caused by the lack of a standard procedure for imperial succession, which was resolved frequently by civil war, crippling the government and weakening
the empire's defenses. "Decadence" has always been a popular
explanation, though in Rome's case the greatest decadence coin-
cided with the greatest power. (Still, here's Richard Nixon on the
subject: "When the great civilizations of the past became prosperous, when they lost the will to go on living and make progress, they fell victims to decadence, which in the long run de-
stroys a culture. The United States is now entering this phase.")
Some of the more preposterous theories explaining the fall of
Rome happen also to be unforgettable. Everybody knows the
one that credits lead poisoning from the pipes used in plumbing.
Another explanation in this class: the onset of widespread impo-
tence caused by hot water in the public baths.

All told, decline-of-Rome explanations fall into two broad categories: either the empire killed itself (internal weaknesses) or
it was killed by something else (external factors). Historians tilt
one way or the other, but they also tend to cite the interplay of
inside and outside forces rather than attributing Rome's demise
to a single simple cause. As in *Murder on the Orient Express,* all the
prime suspects shared in the deed.

The third century was a period of deep crisis, with one bad
emperor after another (the standard obituary: "killed by his own
soldiers") and continual barbarian pressures on the frontiers. The
empire was rescued by Diocletian, a no-nonsense administrator
who built up the legions and organized the state and its taxes

around the need for security. He also created the system known as the tetrarchy, in which power was shared among two co-emperors (East and West) and two subemperors-in-waiting, at a stroke mitigating the succession problem and providing enough leaders to command the various armies. From the late third century onward, emperors spent more and more time away with their legions, fending off trouble and settling in for years in subcapitals like Trier and Sirmium. But Rome was still the leading city of the empire and still the home of the senatorial aristocracy. It still enjoyed extraordinary privileges, and it still extracted great wealth from the rest of the empire.

The sack of the city by Alaric, in 410, was both a physical and a psychic blow. By arrangement the sack was an orderly and not especially bloody affair, its terms spelled out in advance. Alaric demanded all portable wealth, and when a member of the Roman legation asked what the Romans could keep, he replied, "Your lives." Rome's perimeter had not been breached by an enemy for eight hundred years. "The brightest light of the whole world is extinguished," wrote Saint Jerome, the translator of the Bible, in faraway Jerusalem. Over time, wars and sieges took a toll on Rome's water supply—the aqueducts were cut and repaired, then cut and repaired again. The most reliable aqueduct proved to be the Aqua Virgo, which ran mostly underground. You can see an excavated piece of it, and hear the rush of water, under an indie cinema in one of the narrow back streets near the Trevi Fountain. Rome's dwindling population, maybe 80,000 in 600 A.D., withdrew from the fringes and began to cluster near the bend in the Tiber where the Aqua Virgo had its terminus, creating the maze of streets that gives today's city center its medieval character. Rome is a good place to reflect, post-Katrina, on how the failure of infrastructure can shape a community for a thousand years.

Well before its collapse, as skills declined and building materials became harder to get, Rome began to plunder itself for *spo-*

lia, or "spoils," in a way that would seem edgily postmodern if it had been driven by academic theory rather than bitter necessity. The fourth-century Arch of Constantine contains bits and pieces carved in the second century (as the painter Raphael pointed out). Builders of churches and palaces used ancient structures as quarries. The original St. Peter's was supported by columns scavenged from Roman-era buildings; the Renaissance basilica that took its place used marble from the Colosseum. To bolster his claim to an imperial pedigree, Charlemagne imported architectural remains from Rome and Ravenna for his palace at Aachen, and even had some of the stonework fashioned to look like *spolia* from Rome, because it had so much cachet. Meanwhile, in the old imperial capital, blocks of marble and statues by the thousands were burned for lime. A memory of the smoky place where this was done, below the Capitoline Hill, is preserved in the name Via delle Botteghe Oscure, the Street of the Dark Shops. Large tracts of the urban heartland were turned over to pasture, and livestock grazed where Caesars had walked. The process of reversion will not mystify anyone who has visited modern Detroit and noticed trees sprouting on the roofs and ledges of abandoned buildings, or seen how vacant downtown lots are being reclaimed by a hardscrabble urban agriculture.

Such was the spectacle that presented itself to Gibbon on that day in 1764, as he sat looking out over what had once been the Forum and was now the Campo Vaccino, the Field of Cows. In a sentence sagging with judgment and resignation, Gibbon settled on no one cause for the empire's collapse: "Prosperity ripened the principle of decay; the causes of destruction multiplied with the extent of conquest; and, as soon as time or accident had removed the artificial supports, the stupendous fabric yielded to the pressure of its own weight."

A decade later, in 1775, as Gibbon prepared to "oppress the public" (as he put it) with the first volume of *The Decline and*

Fall of the Roman Empire, the British Empire faced a crisis of its own when the thirteen American colonies united in rebellion. Gibbon had celebrated the vaunted freedoms of the Romans, and mourned their loss. He was a friend of Adam Smith, who would become the patron saint of America's economic ideology, his face peering out from the neckties of capitalists. But Gibbon had no sympathy for the "criminal enterprize" of the American Revolution, and as a Tory member of Parliament he supported the campaign of military suppression. When he eventually changed his mind, it was on pragmatic grounds: "I shall scarcely give my consent to exhaust still farther the finest country in the World in the prosecution of a War, from whence no reasonable man entertains any hope of success. It is better to be humbled than ruined." Gibbon never warmed to the rebellion. Visiting Paris during the war, he on one occasion was introduced to the ambassador from Britain's American colonies, Benjamin Franklin. Gibbon was reportedly officious and distant; in a letter he stresses that the meeting was *"by accident."* Sly, funny, self-confident in print, Gibbon in person was profoundly awkward. Franklin is said to have heightened his discomfort by offering to furnish Gibbon with some materials for his *next* book, about the decline and fall of the British Empire.

America's Turn

GIBBON MAY HAVE had no place in his political cosmology for America, but America had a big place for Rome. An obsession with Roman antecedents could hardly have been helped, given the classical education all the Founding Fathers received. My window at the Boston Athenaeum, where I sit right now, looks out over a colonial graveyard, the Old Granary Burying Ground. Every literate person resting there would have known the fabled stories of Rome: the rape of the Sabine women; Horatio at the bridge; the sacred temple geese who gave the alarm and saved the Capitol; Caesar and Brutus on the Ides of March. These were as familiar to them as the D-Day landing or the march on Selma or the Watergate burglary or the convergence of Bill and Monica would be to Americans now.

The educated elite of the thirteen colonies were steeped in the Roman code of *virtus*. In this code there was little room for qualities that today hold pride of place in America. The Founders did not cherish therapeutic notions of self-actualization or self-esteem or "the real me." What mattered was adherence to duty as expressed in outward behavior. The very first Gilbert Stuart portrait of Washington, the so-called Athenaeum portrait, was for a time on display downstairs from where I work. The gaze Washington offers, painted from life, captures no interior spark; it's a serene mask of obligation. In Rome, *virtus* was inextricably bound up with the ideology of Rome's greatness. Here is the Roman legend as one historian sums it up: "A simple, hardy race of peasants, long uncontaminated by the seductive arts and manners of Greece, they held fast to their rustic virtues: sanctity of family life, sobriety of conduct and demeanour, a stern sense of discipline. . . . In consequence of these virtues the Romans achieved their mission, divinely inspired, to rule the world." These were seen to be the values of Rome especially in its republican days, and they were the values the Founding Fathers believed Americans at their best embodied. They're *still* the values we look back on wistfully.

The Roman who epitomized republican ideals was Marcus Porcius Cato, or Cato the Younger (95–46 B.C.), the great-grandson of the Cato who had urged his countrymen toward the Third Punic War with the declaration *"Carthago delenda est!"* — "Carthage must be destroyed!" Cato the Younger was a senator known for eccentric habits, grim austerity, and humorless rectitude — combine Mahatma Gandhi, John the Baptist, and Ralph Nader. He often went about in nothing more than a toga, without shoes or an undergarment. His relations with women were erratic and unhappy. He drank heavily when alone, though his public demeanor was abstemious and his lifestyle bereft of luxury. Cato was not a contented fellow, and was more admired than liked. But his stubborn adherence to Roman virtues and republican principles had no equal. Cato and Julius Caesar were bitter enemies, and Cato tirelessly warned his fellow Romans of Caesar's designs on power. When Caesar eventually triumphed, Cato sought to commit suicide by stabbing himself. He was discovered, and the wounds were bound up, but Cato ripped off the bandages and bled to death; he did not want to give Caesar the satisfaction of sparing his life.

Cato's stand against tyranny echoed down the ages. One of the most popular plays in eighteenth-century America was Joseph Addison's *Cato: A Tragedy in Five Acts.* Today it is nearly unreadable, unrelenting in its uplift and cloying in its nobility, but its tidy verses once spoke as powerfully to Americans as Arthur Miller's *The Crucible* does now. Its influence is stamped indelibly on the rhetoric of the Revolution. Nathan Hale's "I only regret that I have but one life to give for my country" is, in essence, a line from Addison's *Cato.* So is Patrick Henry's "Give me liberty or give me death."

Rome set not just a moral example but a practical one. The Founding Fathers had overthrown a great empire, and now they looked to preimperial Rome for republican political models—in particular for ideas about "checks and balances" that could help preserve that form of government. Cicero had written, "I consider the most effective constitution to be that which is a reasonably blended combination of three forms—kingship, aristocracy, and democracy." Rome's republican government vested these functions in its two consuls, who shared executive power and could serve for only a year; in a senate, made up of the highborn, who served for life; and in the *populus,* the people, who could vote on certain matters. This regime promised a government, Cicero thought, with "an equilibrium like a well-trimmed boat." The checks and balances were taken to such an extreme that if both consuls were leading troops in battle, each took charge on alternate days (a pushmi-pullyu procedure that proved fatal at the Battle of Cannae). Roman precedents were invoked time and again as the drafters framed America's new Constitution. Some of them, of course, were sobering: everyone was aware that Cicero's well-trimmed boat had eventually foundered. Benjamin Franklin's famous remark, when asked as he emerged from Independence Hall what kind of government America now had — "A republic, if you can keep it" — represented a cautionary reference to the unhappy fate of Rome. That, at any rate, is how his listeners at the time would have heard it.

George Washington was the epitome of America's Roman ideal. He was unyielding in his embrace of public virtue, supplementing Roman standards of behavior with the famous "Rules of Civility" written out in his own hand in a schoolboy copybook. ("Every action done in company ought to be done with some sign of respect to those that are present." "When a man does all he can, though it succeed not well, blame not him that did it.") The Romans were obsessed with surveying, an understandable preoccupation for a people expanding into three continents; Washington's first career was as a surveyor — his country was expanding too. Washington knew his Roman history: an invoice survives for an order from an English dealer of "A Groupe of Aeneas carrying his father out of Troy, neatly finished and bronzed with copper, three pounds, three shillings" — a sculpture of Rome's founding legend for the mantelpiece of America's own founder. At Valley Forge, Washington ordered up a production of Addison's *Cato* for his frostbitten army, and attended the performance himself.

After the Revolution, sidestepping suggestions of kingship and returning to his beloved Mount Vernon, Washington was hailed as America's Cincinnatus. Lucius Quinctius Cincinnatus was a Roman of the mid fifth century B.C. who farmed a small plot of land across the Tiber from Rome. As the story goes, Rome was under assault by some neighboring tribes, its army surrounded and on the verge of annihilation. The Romans voted to empower a dictator to lead them out of crisis, and sent word to Cincinnatus, who put aside his plough and came to the city's aid.

> In two days he brought the Romans in sight of the combined armies; he formed his line of battle, and after reminding them what they were to fight for he led them onto the charge with such resistless impetuosity that he obtained a complete victory and gave, as it were, a new life to his country's liberties. Soon as this great work was done, he took an affectionate leave of his gallant army and returned to cultivate his four acres.

That's the conclusion of the Cincinnatus tale as recounted not by some Roman chronicler but by Parson Weems, whose best-selling *Life of Washington* was published soon after Washington's death. In a work commissioned for the Capitol rotunda a few decades later, the sculptor Horatio Greenough produced a massive marble Washington in a classic Roman pose, seated, the toga draped to reveal a bare chest. With his left hand Washington offers his sword back to the people, as Cincinnatus might have done. This sculpture now dominates an entryway at the Smithsonian — it was so heavy that it had to be moved from the Capitol before it fell through the floor. The Cincinnatus reference is probably lost on most visitors: Washington looks like a man in a sauna, asking for a towel.

The Roman ideal ran deep in America for decades. People were so steeped in Cicero that up to the Civil War, the stock form of public presentation was the formal oration. As America began to spread across the continent, and to emerge as an economic power, worries only grew that the country was destined to repeat the Roman story of imperial temptation and humbling decline — worries captured in the painter Thomas Cole's allegorical series *The Course of Empire,* produced in the 1830s. It's not subtle. The series begins with an idyllic depiction of the state of nature, then portrays a moment of imperial sunshine in all its vainglorious fullness, and ends with a painting titled *Desolation.*

Not subtle—but not fantasy, either. It's hardly a stretch to find modern relevance in the example of the Roman Republic, overwhelmed by the consequences of its own growing size and might, and by its perceived national-security needs. In 68 B.C. a pirate attack on Rome's port of Ostia prompted the terrified Romans to cede far-reaching powers to one man, Pompey. There would be no turning back. The need to act boldly and react quickly; to ferret out enemy plans while keeping your own hidden; to show a public face of resolve, concealing doubt and dissent—in Rome, over time, all these mandates produced a change in character. They have done so in America, too. You could point to the expanding power of the presidency relative to the other two branches of government; or to restrictions on personal freedom in exchange for personal safety; or to a culture of secrecy; or to the pervasive influence of the military and the security apparatus. People concerned that America may drift away from a republic and toward a principate, as Rome did, took little comfort from the news, reported and confirmed in the summer of 2004, that the government was "reviewing a proposal" to postpone national elections in the event of some sort of terrorist attack; or from the recent Supreme Court ruling that the police may enter homes without knocking; or from the attorney general's threat to use espionage laws to prosecute reporters for publishing leaks of classified information. Nor have their spirits been lifted by the inroads of a relatively new legal argument known as "unitary executive theory." Among other things, it holds that each branch of the government—not just the Supreme Court—has the right to interpret the Constitution, and it asserts an unprecedented view of the extent of presidential power, including the power to make war without the consent of Congress.

This is not in fact just theory. One concrete result has been the president's practice of appending a "signing statement" to legislation when it comes to him for signature, indicating his intention to enforce the legislation according to his own specific interpretation—if the legislation is enforced at all. Up through the year 2000 American presidents had collectively employed signing statements on about 600 occasions. In the six years since then, the president has added signing statements more than 750 times, on laws pertaining to such matters as the use of torture, whistle-blowing by government employees, the oversight provisions of the Patriot Act, and the obligation of the executive branch to provide Congress with certain kinds of information.

The Roman Empire's penchant for official secrecy was remarked on by the historian Cassius Dio, who complained that because so much had been done behind closed doors, he couldn't get access to materials he needed to write his narrative. He would not have been surprised by the dogged White House effort, which continues in the courts, to conceal the details of the administration's early planning on energy policy and the names of those who participated in it. The Romans had nothing like the technological means that modern America has to create a true surveillance state, but the empire's undercover operatives—the *frumentarii* (who turn up in the video game The Regia) and, later, the *agentes in rebus*—were diligent. In common parlance these operatives were known as the *curiosi*. The philosopher Epictetus, who was born in Rome and knew firsthand the dangers of thinking freely (he was sent into exile), presents a vignette of entrapment in one of his writings: "A soldier, dressed like a civilian, sits down by your side and begins to speak ill of Caesar, and then you, too, just as though you had received from him some guarantee of good faith in the fact that he began the abuse, tell likewise everything you think, and the next thing is—you are led off to prison in chains." Our own *curiosi* have big ears. The National Security Agency, in a program known as Echelon, sifts tens of millions of telephone and data communications every day, searching for any of hundreds of words or phrases that may hint at terrorist activity. Some of them ("White House," "mail bomb" "kill the president") are self-explanatory; others ("Roswell," "blowfish," "Bill Gates") may be counterintuitive. Another program, which bore the name Carnivore until someone started to worry, much too late, about the potential public-relations fallout, essentially conducts wiretaps on e-mail. More recently the national-security apparatus has begun wiretapping the international phone calls of thousands of Americans without legal oversight—on presidential orders, and despite the expressed will of Congress. It has also been collecting the telephone records of tens of millions more.

Then as now, legislatures seem to be the first to go. The Roman Senate remained a millionaire's club and a source of public servants, but it atrophied as a true deliberative body. Foreign policy and war-making power became the sole province of the emperor and his *amici,* his closest advisers. Article 1, Section 8 of the U.S. Constitution states: "The Congress shall have the power to declare war," a power entrusted to the legislature because, as James Madison observed, the temptation to use force would otherwise "be too great for any one man." A modern historian writes, "The debates in the convention, the later writings of delegates to that meeting, and speeches in the state conventions that voted on ratification of the Constitution leave no doubt that the president's title and role as commander in chief gave him no powers that Congress could not define or limit." The last time Congress authorized the use of force through a declaration of war was more than six decades ago, in the aftermath of the attack on Pearl Harbor. Since then the United States has committed large numbers of troops to major combat operations on fourteen occasions, from Korea and Vietnam through Grenada, the Balkans, and two wars in Iraq, but no president has ever sought a true declaration of war. Power has shifted decisively toward the executive, as the executive understands. George

W. Bush made the point this way: "I'm the commander—see, I don't need to explain. . . . I don't feel like I owe anybody an explanation."

The *Omphalos* Syndrome

SOMETHING HAPPENS to imperial capitals, something psychological and, over time, corrosive and incapacitating. It happens when the conviction takes hold that the capital is the source and focal point of reality—that nothing is more important than what happens there, and that no ideas or perceptions are more important than those of its elites. This conviction saturated imperial Rome, as it saturates official Washington—it's the most important trait the capitals share. The conviction is understandable, up to a point. When powerful states are in an expansive phase, the wishes and ideas of the rest of the world seem secondary, inconsequential. In the capital itself, this frame of mind may far outlast the circumstances that produced it, taking on a life of its own that everyone has an interest in perpetuating. It can prove impossible to eradicate fully. In Italy, manhole covers are still stamped with the letters that once appeared on imperial standards and marble monuments: SPQR, for *Senatus Populusque Romanus*—the Senate and the Roman People. Modern Russia can't suppress the reflexes of the old Soviet empire, just as the Soviets couldn't suppress the reflexes of imperial Russia. Britain no longer has much of an empire, but many institutions in London retain a noticeably imperial cast of mind. The spirit of the Raj is not absent from the tone of *The Economist*.

Rome labored under what has been called an *"omphalos* syndrome." The *omphalos*, from the Greek word for "navel," was a stone monument found in a number of ancient cities that supposedly marked the navel, or center, of the world. Rome's own version, the marble Umbilicus Romae, stood prominently in the Forum, right next to the Rostra. The marble facing is now gone, but a circular brick pile remains, to which tourists pay no attention at all, unaware that the entrance to the underworld, to Hades, was once believed to be *right there,* under those very bricks. The term *"omphalos* syndrome" originated in the study of old maps, and describes the tendency of people who "believe themselves to be divinely appointed to the centre of the universe," as one geographer explains, to place themselves in the middle of the maps they draw. The Romans weren't shy about asserting this belief: they drove it home astutely by means of what today would be considered "branding." In the Forum, in the very center of Rome, they erected not only the official Umbilicus but also, a few yards away, a gilded column called the Golden Milestone, where all the empire's roads symbolically converged. Augustus built a sundial the size of a football field in the Campus Martius, using an Egyptian obelisk to cast the shadow. It was dedicated, technically, to a divinity — to the sun — but on the birthday of Augustus, September 23, the obelisk's shadow pointed directly at the Altar of Peace, which celebrated the fruits of his rule. The message was unmistakable, one scholar concludes: "The whole universe now formed part of the new Augustan system."

In the second century A.D., a young rhetorician named Aelius Aristides delivered an oration in the Athenaeum, in Rome, possibly in front of the emperor himself. Aristides was a panegyrist, a court littérateur whose job it was to extol people in power. It is an occupational category that still exists in Washington. (Think of Peggy Noonan on Reagan, Sidney Blumenthal on Clinton, Ron Kessler on Bush, Midge Decter on Rumsfeld.) The description of Rome offered by Aristides embodies the city's self-satisfied outlook: "Here is brought from every land and sea all the crops of the seasons and each land, river, lake, as well as of the arts of the Greeks and the barbarians. . . . Whatever one does not see here is not a thing which has existed or exists." The architect Vitruvius took up the same theme: "Surely then it was a divine intelligence which placed the city of Rome in so perfect and temperate a country, with the intention that she should win the right to rule the world."

Rome, like Washington, was an economically pointless metropolis, a vast importer and consumer of an empire's riches rather than a producer of anything except words and administration (and the pungent cartloads of garbage that left the city every night). The downstream consequences of Rome's gargantuan appetites can be visualized, literally, even today. Take the basic need for building materials. Augustus would claim that he had found Rome a city of brick and left it a city of marble, and the boast was not an idle one. But underneath that marble, and alongside it, mountains of brick remained. Rome's tenement houses, the *insulae*, were faced with brick, and so were the city walls and sewers and aqueducts. The brick was distinctive — square and thin, red-ochre in color. You recognize it everywhere in the ruins of the city, and if you come across it anywhere in the world, whether in York or Paris or Jerusalem, it always means one thing: Rome was here. The brick and lime for the city of Rome had to be baked and kilned, which required massive quantities of charcoal, which in turn required trees. A single burn of a limestone kiln could consume a thousand donkey loads of wood. The forests around Rome were felled, and then the forests beyond those forests. Ground cover gone, the soil washed from the hillsides and into the rivers. At the mouth of the Tiber, the shoreline pushed outward as accretions of soil built up over the centuries. The docks at Ostia had to constantly be extended to remain adjacent to the water, a process clearly visible now in aerial photographs. As the empire came to an end, so did the effort to keep ahead of nature. The original docks are now a mile from the sea, trapped in dry land, separated from shore by striations of silt.

The biggest component of the city's prodigious intake was something called the *annona*, an in-kind tax levied by Rome on everyplace else, and collected in the form of grain, which was used to provide free bread for most of Rome's inhabitants. At its peak the *annona* amounted to 10 million sacks of grain a year. The shipment of the *annona* from Spain, Egypt, and northern Africa to the docks at the Tiber's mouth, and then by barge up the river to Rome, was never-ending, like tanker traffic in the Persian Gulf. Any serious interruption could mean urban unrest to the point of violence. When conspirators wanted to bring down Cleander, the hated chief minister of the equally hated emperor Commodus, they manipulated grain supplies, causing shortages that led to riots. (Commodus "commanded that the head of Cleander should be thrown out to the people," Gibbon writes. "The desired spectacle instantly appeased the tumult.") Eventually the *annona* was expanded beyond grain to include olive oil and wine. The smashed amphorae these liquids came in were tossed in a dump near the Tiber wharves, creating a hill known today as Monte Testaccio, a hundred feet high. Warehouses the size of basilicas existed at every stage of the distribution system, and so, too, did opportunities for pilfering and corruption. If you think of the *annona* as tax revenue, which it was, then the revenue not only accomplished its stated purpose of feeding the city; it also supported large swaths of private-sector activity, from shipping to baking to crime. Some of this activity was encouraged with tax breaks and even grants of citizenship. There was great wealth to be had off government contracts. You can still see today, near the Porta Maggiore, in Rome, the huge marble tomb, in the shape of an old-fashioned bread oven, of a freedman named Marcus Vergilius Eurysaces, who is described by an inscription as "baker, contractor, public servant." So large was the work force required for the baking of bread that people convicted of certain minor crimes were sentenced to hard labor in Rome's bakeries. As the empire began to contract, the *annona* remained the essential lifeline, preserved at all costs. By the fifth century A.D., only the link with North Africa remained unbroken. When Alaric laid siege to Rome, one of his first acts was to send his warriors to seize the docks at Ostia.

To see yourself at the center of everything requires a sense of what "everything" is—a geographic sense, in other words. Americans take such a thing for granted, aware as we are of the location of every place on earth. You can tap into the Global Positioning System with a cell phone. In Roman times geographic knowledge was primitive, though its political uses were not. The Romans called themselves "masters of the *oikumene*"—"masters of the known world"—long before they were able to depict the known world in any reliable way. But once they could, they erected large public maps showing Rome in the literal center, where it obviously had to be. The Roman Empire's vascular system was its network of roads. At regular points along every roadway in Italy marble markers announced to travelers the distance to the center of the world: to Rome. A monument in the capital schematically depicted the deployments of the imperial legions, arrayed in a circle centered on the capital—a precursor, in stone, of the blinking electronic displays of "readiness" in the Pentagon's situation room. Rome's sense of status and privilege would survive long after Roman emperors stopped living there—and, indeed, long after the empire was gone. It survives to this day in the idea of Rome as the Eternal City, and very literally in an ancient pronouncement that occurs every Easter, when the pope from his balcony in the Vatican delivers a homily that begins with the words of address "*Urbi et orbi*—"To the city and to the world."

Rome displayed the attributes of any great capital with more hubris than humility: the overweening self-regard, the presumption that it knew better than others, the surprising ignorance about foreign cultures, the languid arrogance, the competitive displays of wealth—all captured in the writings of Suetonius and Plutarch and Juvenal and others. The city's appetite for the wealth of the conquered knew few limits: as its rule spread to one place after another, a steady traffic in artwork made its way to the center. On occasion Roman grandees would obtain classical sculptures from Greece but replace the heads with their own; think of Rodin's leonine Balzac with the head of a Newt Gingrich or a Joseph Biden. All of this coexisted with a rhetoric of high-mindedness about the duties and burdens of leadership — Rome's "special gift." And the fact is, the rhetoric reflected an undeniable reality: Rome held up its end.

Inside the Bubble

WASHINGTON, TOO, sees leadership as its special gift, though it did not always. For more than a century it was very much a provincial southern town. The Washington of Henry Adams's *Democracy* (1880) is a city of pinched horizons, not the center of anything. There's plenty of religion and corruption and politicking and small-mindedness in the novel, but no sense at all of America on a world stage. Like Rome, Washington changed character suddenly: its Augustan phase began only in the twentieth century and accelerated after the Depression and World War II, spurred by new social ambitions at home and new security obligations abroad. In the eyes of nostalgic proponents of small government, the Rubicon was crossed with the passage of the Sixteenth Amendment, in 1913, giving Washington the unimpeded power to levy an income tax and therefore to spend ever larger amounts of money. In the eyes of those nostalgic for a time when America could hide behind two oceans, the symbolic point of no return is the construction of the Pentagon, rushed to completion in 1943 and still the world's largest office building.

All life in Washington today derives ultimately from the capital's own version of Rome's *annona* — the continuous infusion not of grain and olive oil but of tax revenue and borrowed money. Instead of ships and barges there are banks, 10,000 of them designated for this purpose, which funnel the nation's tax payments to the city. The keystroking civil servants at the federal Financial Management Service, who gather it all in electronically, are Washington's equivalent of the longshoremen at Ostia. The never-ending flow of revenue creates a broad level of affluence that has no real counterpart anywhere else in America. Federal employment may no longer be growing — the federal payroll in the Washington region is about 360,000 — but this is in essence a convenient deceit, to make the size of government seem contained. An even larger number of people in the Washington area — about 400,000 — work for private companies that are doing actual government work; like the baker Eurysaces, they're living directly off the *annona*. H. Ross Perot, the anti-government maverick, made his fortune this way, supplying data systems to an expanding federal government. (He deserves a marble tomb in the shape of an old-fashioned computer punch card.) An additional quarter of a million people in the region feed off government directly or indirectly: the lawyers and lobbyists, the wonks and accountants, the reporters and caterers and limousine drivers and panegyrists, and all the aides and associates whose job it is to function as someone else's brain. Every week a dozen or so pages of the Washington magazine *National Journal* detail the comings and goings of executives in categories like "image makers," "think tanks," "lobby shops," "interest groups" — denizens of the smoked-glass office blocks on K Street, Washington's own Street of the Dark Shops. Washington simply doesn't look like the rest of America. It's richer, better educated, more professional — number 1 in the country in median income, and in the percentage of college graduates, of women in the work force, and of two-earner families. Its professional classes are largely insulated from economic conditions in the rest of the country. As the analyst Joel Garreau has observed, "Only the residents of Washington, reaping the benefits of being at the center of the Imperium, fail to view this as bizarre."

Washingtonians see themselves as the masters of the *oikumene*. When Washington appears in novels these days, it's the Washington that plays the Great Game of foreign affairs and espionage, not the Washington that deals with grubby domestic issues. It's the Washington of Jack Ryan, not Mr. Smith. One analysis of Washington phone books found that listings beginning with the word "international" had increased two and a half times as fast in the second half of the twentieth century as those beginning with "national." Although Washington does not have a marble *omphalos,* the president has his finger on something else that gets everyone's attention: "the button," the one that controls our nuclear arsenal. The president of the United States goes by the acronym POTUS, subliminally evoking potency (from *potens,* the Latin word for "powerful"). For years he was also "the leader of the free world." Now that there is no longer an unfree world, at least officially, the president is simply "the most important man in the most important city in the world." That turn of phrase attaches itself like a limpet to anything in sight. Ads for one Washington bank have described it as "the most important bank in the most important city in the world." The general manager of Washington's power-lunch restaurant The Palm has been called "the most powerful man at the most powerful restaurant in the most powerful city in the world." And hold on: maybe it's not just "the world." As President Bill Clinton prepared for his inauguration in 1993, the *Washington Post* published a four-part series called "A Newcomer's Guide to the Most Important City in the Universe."

The sacred boundary of the city of Rome was known as the *pomerium.* Washington's *pomerium* is of course the Beltway, and "inside the Beltway" has long been conventional argot for the city's special sense of self; "outside the Beltway" means, in effect, "the provinces," "the hinterland." Even time is held captive within Washington's *pomerium:* the atomic master clock at the Naval Observatory, in the heart of the city, defines the meaning of "now" for cell phones and satellites, computers and cruise missiles. When Washington's atomic clock makes an adjustment, even a billionth of a second, time obeys. "Once within the confines of the Capitol complex," the late Meg Greenfield explained, "most people come to accept its standards, live by its rules, honor its imperatives." They also "start referring to the rest of the country (without quite realizing what the term conveys) as 'out there.'" Washington itself, though, is the outlier — an anomaly among American cities. It is not a great cultural capital, just as Rome wasn't; it must import its artists and actors, its rhetoricians, its scholars, its thinkers. Architecturally it is without great distinction, using its past as a kind of *spolia* — erecting new buildings behind old façades (the practice is known as "façodomy"). It's hard to avoid the atmosphere of conspicuous striving — in the slightly overdone formal invitations for every social event, in the leather-and-globe furnishings of offices and studies, in the advertisements for cosmetic surgery in high-end local magazines — and of muted cultural defensiveness. There always seems to be a moment at Washington gatherings when some mildly fortified Oxbridge expatriate begins muttering about how it falls to Britain "to play Athens to your Rome." Washington's wounded riposte would echo that of Julius Caesar in Shaw's *Caesar and Cleopatra:* "What! Rome produces no art! Is peace not an art? Is war not an art? Is government not an art? Is civilization not an art?"

Washington's life-support system is maintained not by one *annona* but by two. The second comes in the form of information — information about everything, public or proprietary, open-source or top-secret, a vast ingathering from all over America and the world to this one place, where it is stored, minced, parceled out, analyzed, palpated, twisted, packaged, shared, and deployed. (Or, in some cases, withheld or destroyed.) William Petty's seventeenth-century *Political Arithmetick,* which used statistics to compare the economic and military resources of England with those of Holland and France, marked a revolution in government. Information has become the brick of the modern state, and the demand for it in Washington is impossible to satisfy. In 2002 the government launched a program known as Total Information Awareness, whose name sums up its aims, to be pursued through electronic and other means. Justified on national-security grounds, and run out of the Defense Department, the program encountered opposition because of privacy concerns, and was said to have been shut down. But many of its components continue under other names. As all roads once led to Rome, all computer trunk lines lead to Washington. In locations throughout the city — on Capitol Hill, in the White House, in the offices of defense contractors and political lawyers, there are secure redoubts known as "SCIF rooms"; the acronym, pronounced "skiff," stands for "sensitive compartmentalized information facility." These are special sites, protected against eavesdropping, where information of utmost gravity can be imparted to those select few with clearance to receive it. One participant in SCIF discussions says, "Those of us who've been in those rooms long to be in them again."

In the Ellipse just south of the White House stands a granite Zero Milestone, intended to be Washington's version of Rome's Golden Milestone, the symbolic central reference point from which all things are measured. Well, it isn't our central reference point at all — no one has ever heard of it, though you could argue that modern America began on this very spot. This was the place from which Lieutenant Colonel Dwight D. Eisenhower, in 1919, set out to lead the army's "transcontinental motor convoy" across America. By the time Eisenhower reached San Francisco, sixty-two days later, he understood that America needed what Rome had possessed, a network of good public roads. When he became president, he created the interstate highway system. Tourists pay no attention to the Zero Milestone at all, and yet our own descent into hell started *right there*.

Washington's real focal point is provided by the Washington press corps, which provides the magnifying lens through which the capital is seen. Increasingly, the national and international news cycles are set by journalists and media executives inside the

Beltway. If you count by the minute, as much as 30 percent of the nightly network news is datelined Washington, and most of that coverage is built around official sources. Of the 414 stories on Iraq broadcast on the three major networks from September of 2002 through February of 2003 (that is, during the run-up to the war), 380 were reported from the White House, the Pentagon, or the State Department — as opposed to, for instance, any European capitals, or Middle Eastern capitals, or the United Nations, or Iraq itself. The weekly Friday-evening and Sunday-morning political talk shows begin with a musical flourish that evokes imperial trumpets, and the words "From Washington . . ." The analysts offer what James Wolcott has called a "luxury-skybox view" of national affairs:

> Assurance fluffs up their every pronouncement, because they have permanent thrones . . . Not having to answer to angry constituents, they make everything sound easy. They dispatch imaginary troops overseas as if snapping their fingers for a taxi. Welfare cuts? No problem. Slash government payrolls? Make it so.

"The week" — meaning the political week in Washington, and in essence meaning the president's week — has long been the basic formal unit of journalistic time (has POTUS had a good week? a bad week?), although with the Internet and round-the-clock cable news, the half-life of particular stories keeps getting shorter and shorter. The ephemeral nature of "importance" in the capital is symbolized by the "Zeitgeist Checklist," in the *Washington Post*, which rates the urgency of various issues — immigration, same-sex marriage, Iraq — as if they were items on a bestseller list (current rank, last week's rank, weeks on list). It's done skillfully, with practiced irony. But it works only because in this instance the ironic and the real overlap completely.

In Rome at its height all social and political life was derived ultimately from the imperial court; and only through access to the court could those men of influence known as *suffragatores* control jobs and resources. In their day the names of *suffragatores* like Libanius and Themistus and Fronto were as expeditious as those of Vernon Jordan and Robert Barnett in our own. Money fuels the system, of course — no surprise there. In the days of the republic, when Rome still had functioning electoral elements, the inexorable creep of lobbying and vote-buying induced pathetic fits of "campaign-finance reform," such as placing limits on the numbers of people a candidate could invite to banquets, on how much a candidate could spend on food and drink, and on where the banquets could be held. As Washington would discover to its own great relief, the only effect of such measures was to prompt the discovery of new loopholes.

The degree to which the world inside the *pomerium* has become a hermetically sealed system is taken for granted. Jimmy Carter thought of Washington as an island (and was thrown off it). Eisenhower complained that everyone in Washington "has been too long away from home." The occasional acknowledgment of how isolating Washington is changes nothing. *Newsweek* ran a cover story in 2005 called "Bush in the Bubble," because the president and his advisers seemed to be living inside a membrane that kept certain viewpoints in and certain realities out. The description fits any recent administration, though this one more than most. Documents pertaining to Vice President Dick Cheney's travel requirements became public in 2006, revealing that when he entered a new hotel room he wanted all the television sets already turned on and tuned in to the ideologically congenial Fox News. (With the hiring last year of the Fox News anchor Tony Snow as the White House spokesman, the circuit has been closed.) To keep the membrane in good repair, the vice president has required crowds behind rope lines, waiting to shake his hand, to cleanse themselves with antibacterial gel.

Pliny the Elder describes the marvels of Rome's drainage system, the sewers leading down from the seven hills into the central Cloaca Maxima, which another Roman observer once called "the receptacle of all the off-scourings of the city." Washington now drains into the blogosphere, another engineering marvel. In the larger scheme of things, how important were the personnel changes at the White House last year, which removed Scott Mc-Clellan as press secretary and took away one of Karl Rove's jobs? They came at a moment when the price of oil had reached $75 a barrel, Iran was pursuing plans to build a nuclear bomb, and the war in Iraq offered a prospect of perpetual carnage, but for a week the swirl of opinion in Washington blogs — scores of sites, hundreds of links, tens of thousands of words every day — centered mainly on those White House personnel changes, and who won and who lost, and how prescient (or asleep) the bloggers themselves had been. Archaeologists excavating the drains of the Colosseum have found the bones of exotic animals and the remains of stadium food. Archaeologists dredging the Washington blogosphere will discover dense strata of self-reference: "as I said this morning . . . ," "as I predicted last week . . . ," "I'll say it again . . . ," "to repeat . . ."

Within any closed, insular system, the competitive pressure for status becomes intense. Edward Gibbon, in a typically tart moment, took note of Roman officialdom's taste for fancy forms of address: "They contend with each other in the empty vanity of titles and surnames, and curiously select the most lofty and sonorous appellations . . . which may impress the ears of the vulgar with astonishment and respect." Washington titles do not approach the grandeur sought by North Korea's Kim Jong Il ("Saint of All Saints," "Lodestar of the Twenty-first Century"). But consider this: during the Kennedy administration only twenty-nine people held the coveted title of "assistant," "deputy assistant," or "special assistant" to the president; by the time Bill Clinton left office, there were 141 such people. Unabashedly ambitious (Cicero maintained that the quest for *gloria* explained everything), the Romans spelled out their achievements with painstaking care in autobiographical *commentarii* when alive, and in detailed *elogia* by allies when dead; the self-serving, score-settling "Washington memoir" has a long pedigree. In his letters you can watch Cicero hire buyers and decorators to make his villa outside Rome into a statement of good taste and great influence. He would have been an avid consumer of the American capital's glossy shelter and real-estate magazines. ("Its spectacular glass and columned façade speaks of power and professionalism," says one ad for Washington office space.) The Washington world of public-relations "handlers" and of "strategic communications" had a counterpart in Rome, where applause in the law courts could be bought and sold at standard rates. When a Roman was awarded an official triumph through the streets of the capital — the pinnacle of public achievement — painted renderings of his deeds were carried along in procession: a mobile version of that Washington fixture, the "I love me" wall, with its photographs of the *triumphator* gripping hands with the mighty. The quintessential Washington text may in fact be a Roman one, Cato the Elder's self-promotional speech titled "On His Own Virtues."

The *omphalos* syndrome is not just a curiosity — it leads to isolation and a view of yourself and the world that can be sharply at odds with the true state of affairs. Rome actually had more insulation against the consequences than Washington does. Roman emperors traveled continually, whether to wage wars or just to see the empire for themselves; they could be absent from Rome for years at a time. Moreover, for all his power, a Roman emperor could be an oddly passive figure. In his magisterial study of how Roman emperors did their jobs, the historian Fergus Millar describes what might be called an "in-box imperium." The majority of an emperor's time was spent simply considering petitions and then rendering decisions — often in person, with supplicants from all over the empire standing before him to state their business. His time was not spent dreaming up social programs, defending civil rights, or reinventing government; any talk of pressing toward a New Frontier would have been meant literally — adding territory. Finally, the empire's far-flung parts were run by capable proconsuls of high stature, their autonomy enhanced by great distance and poor communications. The Roman mindset — center of the world! — might be a palpable reality, but in practice the nature of government put limits on its scope.

In Washington it's exactly the opposite: the nature of American government amplifies the mindset of the capital. A president is deemed a failure if he is not pushing an activist agenda. He is therefore wary of being seen as "detached," wants to be seen as "hands-on." The president spends most of his time in the capital, and even on his many short trips he remains largely isolated from ordinary people. The machinery of government centered on Washington — hundreds of agencies, millions of workers — had no counterpart in Rome. The machinery is there to be used, and a president has access to all of it. Modern communications ensure that no job is beyond potential presidential supervision, even when decentralization and autonomy might be all to the good. Lyndon Johnson personally selected bombing targets in Vietnam. The commanders of the failed military rescue mission in Iran, in 1980, had to check with Jimmy Carter's White House by radio every step of the way. With its vast political databases a modern White House can tailor specific messages to individual households everywhere, as easily as VISA or Comcast can, reflecting a Washington presumption that "out there" is subject to manipulation from the center.

In America, then, as a practical matter, the workings of Washington encourage the idea that the world is small, that society is malleable, and that the capital's stance is paramount. All things begin in the capital, the prime mover of all change. You see traces of this idea in everything from the War on Poverty, in the 1960s, to the Clinton health-insurance plan, in the 1990s.

In foreign policy the idea makes itself felt as resistance to multi-lateral arrangements (such as the treaties to reduce atmospheric pollution and to ban land mines and antiballistic missiles) and as faith in unilateral action (such as pre-emptive war). Across the board it fosters the conviction that assertions of will can trump assessments of reality: the world is the way we *say* it is. Thus, in the most recent federal budget, $20 million has been set aside for an eventual "day of celebration" marking American victory in Iraq and Afghanistan.

A Roman moment captures the spirit: in 476 A.D., not long after the last emperor was deposed and the empire in the West had come to an end, the Roman Senate ordered new coins to be struck. The coins bore the legend *"Roma Invicta"* — "Rome Unconquered."

2

THE LEGIONS
When Power Meets Reality

Valens was overjoyed at the prospect of so many recruits. He
reasoned that if the Goths were integrated into his army it
would give him an overwhelming force.
— Ammianus Marcellinus, *Roman History*

I would go further and offer citizenship to anyone, anywhere
on the planet, willing to serve a set term in the U.S. military.
— the columnist Max Boot

BAGRAM AIR BASE, in Afghanistan, is located on a sere
plain beneath snowcapped spurs of the Hindu Kush,
about thirty miles north of Kabul. Alexander the Great founded
a city, Alexandria of the Caucasus, a few miles north of the air
base. Even closer to it Alexander built several forts. Under other
circumstances archaeologists would be busy here. In the past,
Bagram has yielded glassware and bronzes from as far away as
imperial Rome. Its carved ivories, looted from Afghanistan's national museum during the country's three decades of war, are
prized items on the international black market. If archaeological
work is not much in evidence these days, it is because the sites
are overlaid with some of the world's most extensive minefields,
a legacy of the Russian occupation, from 1979 to 1988. There's
a strategic rationale for this acute level of military interest: Bagram lies on a stretch of the ancient Silk Road. It guards the

southern approach to the Khawak Pass, a route through forbidding mountains traveled centuries ago by Tamerlane and Genghis Khan as they made their way south into India. Bagram was also a redoubt held by the Northern Alliance and its American advisers in 2001, and it was from there that they pushed toward Kabul to drive out the Taliban.

Bagram today is an outpost of American, not Hellenic, civilization. Alexander's city supported a population of some 4,000; Bagram Air Base supports a population of more than 5,000. The base perimeter, nine miles around, is ringed not with walls of stone or mud but with chain-link fencing and concertina wire and arrays of bright lights and electronic sensors. The Russians may have built the airfield runway and the control tower, but if you subtracted the particularities of view and climate you could be on almost any American military base in almost any part of the world. There are the same orderly rows of prefabricated dwellings; the same stacks of shipping containers; the same giant bladders of water and fuel. There is a mini–Main Street with American-style stores, and a small hospital offering some of the best medical care in Afghanistan. There is a precinct where the hundreds of civilian contractors live, doing the cooking and cleaning for the troops; and the checkpoint guarding and facility constructing; and, outside the perimeter, in greater Afghanistan, a lot of the nation building. In your spare time you can take college courses or watch satellite TV, and for $50 a month you can get Internet access. A dozen varieties of religious service are offered at a containerized facility, the Enduring Freedom Chapel, which comes stocked with prayer rugs and chalices and menorahs, and with sacred books covered in a camouflage pattern. The chapel is a shrine to America's civil religion as well: naturalization ceremonies have been held here for American personnel in Afghanistan who were not citizens when they joined the armed forces, as increasing numbers of recruits are not. Well-meaning celebrities from the United States come and go: Robin Williams, Roger

Clemens, Henry Rollins, Drew Carey, Al Franken, the Dallas Cowboys cheerleaders. And then, of course, there is the war-fighting aspect of the air base — the fighter jets and helicopters, the shuttling transport planes, the bristling arsenal.

The uniformity from American base to American base is a function not just of military culture but of military spending. A single company, Kellogg Brown & Root, until recently a sub-sidiary of Halliburton, is responsible for most of the physical infrastructure and maintenance, and many of the services, and sometimes even the security, at America's bases in Afghanistan and Iraq and the Balkans and elsewhere around the world. An-other group, the Army and Air Force Exchange Service, oper-ates more than a thousand fast-food restaurants on U.S. military bases worldwide. The Burger King outlet it runs out of a trailer at the Baghdad International Airport is one of the ten busiest Burger Kings in the world. The Exchange Service oversees de-partment stores and movie theaters. It issues its own credit card, the Military Star Card, and since 9/11 it has absorbed the Star Card debt of any member of the armed forces killed in action. The Army and Air Force Exchange Service ensures that Baskin-Robbins and Mountain Dew, Levi's and L'Oréal, will follow Old Glory wherever it goes.

Robert D. Kaplan, an American correspondent who has been in and out of scores of American bases around the world over several decades, writes of a recent stay at Bagram:

It takes so little to reproduce a material culture, I thought. HESCO barriers, a bit of cheap carpentry, rows of portable toilets, Armed Forces radio, two weight rooms with Sheryl Crow CDs playing in the background, and half a dozen chow halls serving fried chicken, collard greens, and Snapple and Gatorade — and boom! You've got the United States. Or at least a particular country-slash-southern-slash-working-class version of it.

Or, to import this observation into another cultural context: it takes so little to reproduce a material culture — a bit of cheap jewelry, rows of stone toilets, mass-produced plates and bowls from southern Gaul, chow halls serving imported wine and olive oil, styluses and wax writing tablets for letters to friends and family, bathhouses with heated floors, carved effigies and altars honoring the familiar gods from back home — and boom! You've got the Roman Empire. Or at least a particular north-of-Britain-slash-barbarian-auxiliary version of it.

That is the thought I took away from Vindolanda, a Roman camp in Northumberland that archaeologists have been excavating for nearly seventy-five years. The landscape itself no longer has any sort of military feel, although the sonic signature of British Harrier jets, flying low in pairs from a base nearby, sometimes breaks the silence. Vindolanda is otherwise a Constable painting, occupying a grassy hilltop with a panorama of rolling moors, craggy escarpments, and grazing cattle, and within easy reach of friendly pubs. Through it runs the Stanegate, the old Roman supply road that parallels the River Tyne across the neck of Britain. A weathered Roman mile marker stands nearby, hard to miss. The emperor Hadrian surely noticed it; he spent time in this very spot, overseeing the construction of the wall, a little to the north, that bears his name. Vindolanda was initially fortified around 85 A.D., and it was occupied, more or less peacefully, for the next three centuries, until the Romans decamped from Britain altogether, pulling back their legions to deal with crises elsewhere. Vindolanda's moats and ditches, which served as casual trash pits, have so far yielded more than 5,000 Roman boots and shoes, along with many other things — coins and jewelry, plates and bowls, hides and headgear, knives and armor. Today the excavations are presided over by Robin Birley, a bluff man with fists like oaken burls, who loves the work in part because he has no idea what will come out of the ground next ("except shoes").

Who was stationed at Vindolanda? For a time, mostly Batavians and Tungrians — Latin-speaking auxiliaries of barbarian stock originally from what are now the Low Countries. These cohorts once guarded their native regions, but they had proved unreliable during the tumultuous "year of four emperors," 69 A.D., a period of civil war, political assassination, and barbarian revolt that followed the suicide of Nero (and that, if nothing else, helps give some perspective to *Bush* v. *Gore*). Henceforward it became Roman policy not to station barbarian auxiliaries on their native soil. And so the Batavians and the Tungrians found themselves transferred to the north of Britain. Their bitter anthem could well have been some version of W. H. Auden's "Roman Wall Blues": "Over the heather the wet wind blows, / I've lice in my tunic and a cold in my nose." The poem goes on: "The mist creeps over the hard grey stone, / My girl's in Tungria; I sleep alone."

What is remarkable about Vindolanda is that it is not remarkable at all — it precisely resembles the Roman bases that exist by the hundreds all over the empire. The fortified playing-card shape is the same, and the interior street pattern, and the orderly rows of barracks, and the types of amenities, and the variety of imported goods. Fly over the Syrian Desert when the sun is low, or along the Dead Sea, or across the emptiness of Libya and Tunisia, and you will get a sense from the bas-relief rectangles — on the edge of a deep wadi, in the cup of a high valley, on the flat of a lunar plain — of the vast extent of Rome's military apparatus, and of its cookie-cutter infrastructure. For the sake of morale at these lonely outposts, performers were brought in from afar to amuse the troops; even Trajan, not remembered as a soft touch, imported actors from Rome to entertain the legions in Syria. Courier services linked the bases to one another, and to the empire's cities and towns. The messages exchanged have hardly changed in 2,000 years.

We know this for sure because of Vindolanda. A chance conjunction of British climate and Roman building methods created anaerobic conditions deep below the surface at the site, preserving many items that would otherwise have decomposed. Cloth has been pulled from the muck still bearing the *clavus,* the broad purple stripe of the Roman aristocracy, an ancient version of the red power tie. Most important by far have been the well over a thousand delicate bits of writing, scrawled in ink on fragments of paper or etched on thin wooden tablets coated with wax. Some of them contain details (such as personal names) that are obviously specific to a particular time: "Octavius to his brother Candidus, greetings." Many offer specifics about life on an imperial frontier: "Nails for boots, number 100" and "As for the 100 pounds of sinew from Marinus—I will settle up" and "I ask you . . . to grant me leave." In one fragment a Latin epithet —*"Brittunculi"*—displays the figure of speech known as the contemptuous diminutive, revealing a certain Roman disrespect toward the locals: "pathetic little Brits" gets the idea across. But much of the Vindolanda writing—letters received, or drafts of letters later recopied and sent—is ageless, indistinguishable from e-mail sent home by American forces in Iraq and Afghanistan.

Here's a jumbled selection from U.S. military e-mail traffic and the Vindolanda fragments: "I am surprised that you have not written anything back to me for such a long time." "I am beginning to wonder if you are mad at me or what?" "Be good for Grandma. Daddy loves you and will see you soon." "Send me some cash as soon as possible." "Still no hope in sight on when we will get the hell out of here, so I will continue to cross my fingers and do the Goat Dance." "While I am writing this to you, I am making the bed warm." "I have sent you pairs of socks, two pairs of sandals, and two sets of underwear." "For the celebration of my birthday, I ask you warmly to come to us." "Well, on that happy note: happy birthday to you."

The Grand Tour

NO COMPARISON involving Rome and America receives more attention or is more frequently invoked than the military comparison. America's "seemingly imperial power," in the words of the Harvard political scientist Joseph Nye, "is more dominant perhaps than any other since the Roman Empire." Once, when visiting American forces on the front lines of the Cold War, in the forests of old Germania, Senator Daniel Patrick Moynihan — not a "seemingly" or "perhaps" kind of fellow — couldn't help himself: "This is the stuff of Roman legions!" he exclaimed. And in a certain brute sense the comparison has something going for it: Rome and America are both big dogs. The Romans dominated most of the world they knew about, the *oikumene* (though not all of it); the Roman Empire is the only entity in history that ever controlled the entire Mediterranean coastline. For its part, the United States not only has no military rival; its level of annual spending on the military is equivalent to that of the next fifteen or so countries combined, and its stated ambition is to achieve "full-spectrum dominance." We're the number 1 economic power in the world, and we provide a security umbrella for the number 2 power (Japan) and the number 3 power (Germany). With alliances spanning the continents and a navy patrolling the seas, America shoulders some of the defense responsibility for billions of people. It's a prodigious undertaking.

What happens if you can't keep it up? The pressures come from more than one direction. There is the sheer managerial challenge of policing whatever you conceive to be "the world." It's difficult when, like Rome, you don't have modern communications and transport — but in some ways just as difficult when you do, given that your enemies may also have these technologies. For both Rome and America, the nature of the perceived security threat keeps changing. Is it from outside powers of considerable might and sophistication (the Persian Empire, the Third Reich)? Is it from internal rebellions? The tribes on your borders? Religious zealots? International terrorists? Your own generals? Meanwhile, keeping vast armies permanently in the field demands an enormous outlay of treasure. The domestic budget of imperial Rome was a meager thing by comparison with that of modern America, with its array of entitlements and services, but Rome often had to squeeze its people hard to extract the money it needed for national security, and it continually deval-ued its currency to make ends meet. (Deficit spending as we know it did not yet exist.) America achieves the same result by borrowing trillions of dollars, going ever more deeply into debt while trying not to worry about the many serious national needs it's simply ignoring. There's one military resource that not even an emperor can just conjure into existence, however: flesh-and-blood human beings. Manpower shortages become a big prob-lem for both Rome and America—and both arrive at the same unsatisfactory solution.

A few years ago, I had a chance to spend some time visiting a dozen or so of America's major military installations: among them, the naval base at Coronado Island, in San Diego; the Ma-rine Corps base at Camp Lejeune, North Carolina; the Army Ranger School at Fort Bragg, also in North Carolina; and the Air Force facility in Colorado Springs known as the U.S. Space Com-mand, which oversees America's high frontier. I was flown onto and off an aircraft carrier, and in a clearing amid scrub pines watched with night-vision goggles as paratroopers by the hun-dreds floated down like green jellyfish in a murky sea. It was all designed to leave a lasting impression—and it did. The Romans obviously had no air force, much less a *praefectura caeli*, a space command, to move satellites around, but a tour of its military facilities would have seemed remarkably similar and no less as-tonishing. The tour might have started with the great naval base at Misenum, north of Naples, where the main portion of the Ro-man fleet was based; sections of the ancient complex remain vis-ible under the wisteria and honeysuckle in what is today partly a beach town and partly still a *zona militare*. Perhaps the next stop would have been the Castra Peregrinorum, in Rome, whose foundations lie underneath what is now the elegant little church of San Stefano Rotondo. This was the headquarters of the *fru-

mentarii, the military agents who originally managed the all-important grain supply, and who gradually came to serve as the emperor's couriers and as his eyes and ears — his snoops, often his hit men. (The name, which means "purveyors of grain," must have come to seem like a bizarre euphemism, the way using "youth services" for "jail" does today.) Where next? Maybe to one of the imperial stud farms in Spain or Thrace or Cappadocia. The tens of thousands of horses required by the Roman cavalry had to be bred and trained, taught to wheel and turn, to jump and swim. Next: a tour of the *fabricae* — the artillery factory at Augustodunum, say, or the arrow factory at Concordia, or the bow factory at Ticinum, or the shield factory at Marcianopolis, or maybe the factories for cavalry armor at Antioch and Nicomedia. The Romans didn't have Henry Ford–style mass production, but they did have factories and they did have a military-industrial complex. Three dozen of these *fabricae* were located strategically throughout the empire to provide the legions with armaments; some of them were general suppliers of swords and shields. Others were far more specialized: the Colt Industries and Lockheed Martin and Point Blank Body Armor of their day.

There are a host of differences between the American and the Roman fighting forces. In terms of technology, for instance, America enjoys a huge advantage over any enemy — too great to be quantifiable, and far greater than Rome usually possessed. Rome generally had a qualitative upper hand, though sometimes, as against Parthia's heavily armored cavalry, the fabled cataphracts, it was at a disadvantage. But two great similarities stand out.

One is simply the logistical capability of the two armies. If it were a private corporation, the Defense Logistics Agency, which buys and tracks all of the U.S. military's basic supplies, would rank number 50 on the Fortune 500 list, right above Intel. The agency operates the Defense Supply Center, in Philadelphia, which grew out of the Schuylkill Arsenal, the group that outfitted Lewis and Clark. In one recent year the supply center distributed 2.5 million pairs of new boots, 4,000 different kinds of prescription drugs, and 285,000 bottles of sunscreen. The mobilization for an overseas posting of an American brigade is a thing of organizational beauty, a martial symphony of cranes and flatcars, scored in bar codes and conducted by sergeants with clipboards. Even a small deployment — 4,000 people to, say, the Balkans, the old Roman diocese of Pannonia — is a taxing and complicated event, unacknowledged by the op-ed strategists who treat force projection as if it were essentially a very large game of Risk. And yet the U.S. military turns impressive feats into the stuff of routine: an Army Ranger battalion, 600 strong, ready to deploy in eighteen hours; a 3,500-man armored Stryker brigade, capable of moving from America to Anywhere in ninety-six hours. This is the grist of propaganda — but it's not *merely* propaganda.

In terms of logistics, the world had never before seen anything like the Romans — and once the Romans were gone it would not again see a fighting force as complex and well managed until the rise in the seventeenth and eighteenth centuries of the globe-circling Royal Navy, with its supremely competent and logistically sophisticated Victualling Board. It has been estimated that a single Roman legion — about 5,000 men — required a thousand horses and other pack animals just to transport baggage, plus more to haul food and fodder. The U.S. military is often criticized for its enormous fuel consumption. The Bradley Fighting Vehicle, for instance, gets only two miles to a gallon; the M1 Abrams tank gets only one. A mobile Roman army was also in the energy business. In meat alone a single legion in a single day ate the equivalent of a dozen oxen or ten dozen sheep. Vast herds were driven behind the fighting men. An army on the move might try to live off the land, and to that end a sickle was standard equipment for every soldier. But a stationary army on the Rhine or the Danube needed, and got, a steady stream of supplies from throughout the empire. Amphorae from the Mediterranean world turn up all over the German frontier. Excavated Roman latrines in northern Britain yield the seeds of figs from the Middle East.

The road system was the structural underpinning of Rome's power, and engineers traveled with the legions to repair bridges and roads or build new ones. They brought along everything they would need. Excavations at the site of one legionary fortress, guarding access to the Highlands of Scotland, uncovered a cache of 750,000 nails. The ancient sources brim with tales of logistical and engineering feats, recounted as matter-of-factly as news of an American airlift of medical supplies to Fiji or Belize. In 55 B.C. Julius Caesar, facing the Suevi across the Rhine, built a wooden bridge so that his legions could march into Germany, deeming it "unworthy of his own and the Romans' dignity to cross in boats." Then, after marching back, he ordered the bridge destroyed, as if to demonstrate to the barbarians that such marvels were as nothing to Romans. During his own, later campaigns against the tribes across the Rhine, the Roman general Germanicus twice built from scratch a new riverine fleet, a thousand ships strong.

The Romans excelled at training—the second great similarity. Americans are accustomed to thinking of their armed forces as "the best trained and best equipped" in the world (to use the stock locution of presidents). Not long ago, when critics pointed out that almost no units of the Iraqi army approached anything like American standards of readiness, administration officials sought to soften the charge by pointing out that most NATO units don't approach American readiness either—it's just too high. In the imperial military's prime, Roman skill, esprit, and group identity rivaled that of Americans. Arduous drilling in peacetime was continual. A military handbook from the second century A.D. explains, "Trained by sweating, puffing, and panting, exposed to summer heat and bitter cold under an open sky, the soldiers become accustomed to the future hardship of real fighting." The Roman-Jewish historian Josephus, who witnessed and chronicled the brutal Roman campaign against the Jews in the first century A.D., offered an assessment of Roman conditioning: "One would not be wrong in saying that their manoeuvres are like bloodless battles, and their battles bloodstained manoeuvres." The legionaries remained tightly knit and self-contained even when all was quiet on the far frontiers; careers in the military spanned generations, as in the American South. "All professional armies form, to some extent, closed communities with their own customs and standards of behavior," one historian has observed, "but the Roman army, more than most, effectively formed its own society."

Americans are familiar with places like Twentynine Palms and Fayetteville, where major bases have given rise to sprawling "civilian" towns beyond the gates, with their tattoo parlors and fast-food boulevards, their ticky-tacky apartments and strip-mall evangelical ministries. Roman garrisons were much the same; there was a civilian *vicus*, or settlement, around Vindolanda, as there was around any significant outpost. (A tent city of merchants even grew up next to the Roman army laying siege to Masada, out in the desert.) And if a soldier was lucky enough to live until retirement—typically after service of twenty-five years —then he, like his American counterpart, might very well settle alongside other retired comrades, collecting *praemia*, or bonuses, in the form of land or money. York, Lincoln, and Gloucester were all *coloniae* for veterans—the San Antonio and San Diego and Norfolk of their day. The word *colonia* survives—hidden in plain sight—in the name of one well-known German city: Cologne. In modern America it is not unusual for a military professional to put in twenty years and then, having earned retirement and a comfortable pension, leverage that experience into a private-sector job with one of the defense companies. Consider this Roman tomb inscription: "With good fortune, Flavius Zenis lived for fifty years; having served in Legion XI Claudia, he enrolled in the *fabrica* of Marcianopolis for twenty years' service as a *centenarius*." He went to work, in other words, as an executive at the shield factory—he went to work for the equivalent of Point Blank Body Armor.

Too Big, and Too Small

ONE OF THE POINTS of comparison often suggested between Rome and America has to do with the notion of "imperial overstretch"—the idea that one's security needs, military obligations, and globalist desires increasingly outstrip the resources available to satisfy them. This suggestion frequently carries the implication that the strategic situations of America and Rome are in essential ways comparable. Three decades ago Edward Luttwak began his influential monograph *The Grand Strategy of the Roman Empire* by citing "Roman imperial statecraft" and "the fundamentals of Roman strategy" as touchstones for "our own disordered times." He was taken to task by one prominent reviewer, who called Luttwak a creature of the "lessons of history" school, and went on: "By suggesting that his reader should see similarities between the strategic problems of Jimmy Carter and those of Marcus Aurelius, he turns this interesting—if flawed—history into an enigmatic allegory."

Well, any argument can be taken too far, though Luttwak, in fairness, invoked the contemporary comparison only in passing. But some parallels do stand out. Both Rome and America develop high-maintenance militaries far more skilled and expensive than those of any competitor. Both put great stock in battlefield ferocity—"shock and awe"—and both regard "making a commitment" as sacrosanct, for reasons of honor and credibility. Both Rome and America, at their most adroit, have understood the influence of "soft" power—the power projected on the wings of language, culture, know-how, luxury goods. Both Rome and America, again at their most adroit, benefit from the

psychological impact of military strength: how the mere percep-
tion of power begets additional power, and makes unnecessary
the actual use of force. In 1995, at an air base in Ohio, as bellig-
erents from the Balkans argued stubbornly over the details of
the Dayton Accords, American negotiators fostered the coopera-
tion of the Serbian leader, Slobodan Milosevic, by seating him
directly under a U.S. cruise missile. Once, to encourage a spirit
of mature reflection among a delegation of barbarians, the emperor Hadrian ordered a unit of mounted cavalry — in full armor,
in perfect formation — to swim across the Danube and back. The
barbarians, writes Cassius Dio, "stood in terror of the Romans,
and turning their attention to their own affairs, they employed
Hadrian as an arbitrator of their differences."

At the most expansive strategic level of all, that of historic
purpose, both Rome and America have considered their way to
be the world's way. As early as the 1830s Alexis de Tocqueville
described America as "proceeding with ease and celerity along
a path to which no limit can be perceived." America believes
that the Western creed of political democracy and free-market
economics is applicable to everyone. That these ideas no longer have significant competition — and won't ever again — is the
whole point of Francis Fukuyama's argument about our having
reached "the end of history" (though Fukuyama did not argue
that every place on the planet would get there immediately, or
even soon, much less be force-marched at the point of a gun).
Virgil, a poet in a political scientist's role, gave the Romans an
explicitly imperial destiny. So did Pliny the Elder, in an evocation of Rome that combines equal measures of Ronald Reagan and Emma Lazarus: "A land chosen by divine providence
to unify empires so disparate and races so manifold; to bring to
common concord so many rough, discordant voices; to give culture to mankind; to become, in short, the whole world's homeland."

At the same time, the differences are considerable. Nuclear
annihilation was not a source of concern (or an option) for the

Romans. And then there's the empire question. The larger goal of American foreign policy for half a century has been to create a stable international regime that gives America's version of free-market capitalism as much latitude as possible. According to this logic, as one analyst observes, America must "protect the interests of virtually all potential great powers so that they need not acquire the capability to protect themselves" — because the possession of such a capability has in the past always led to trouble. Ask the experts whether this form of American hegemony is tantamount to empire, and they'll never stop arguing; it doesn't quite fit either of the classic models, the "territorial empire" or the "overseas empire." Whether it's sustainable remains to be seen. But America, unlike Rome, hasn't set out very often on deliberate missions of imperialistic conquest. It still prefers (as Rome did at first) to leverage the capacities of client states and the local ruling class, as opposed to administering distant domains from the center — an approach that the historian Charles Maier describes as a "consensual empire" or an "empire of invitation." American presidents don't embark on military adventures mainly for purposes of personal glory, as Caesar did in Gaul and Claudius did in Britain. Only metaphorically does America exact tribute. It tends not to pillage other countries, at least not in the form of a destructive rampage, though in economic terms it may suck them dry. It has treated many of those it defeated in war (Germany, Japan) in a manner the Romans could not have imagined. There was no rehabilitative Cato Plan for a conquered Carthage.

The default presentation of Roman "grand strategy" during the empire remains Edward Luttwak's. It has plenty of critics, who take issue with one aspect or another, and there are many who support Luttwak's general scenario but dispute how conscious the strategy was — in other words, whether it was actually "grand." Still, friend and foe alike keep returning to Luttwak as a baseline, the place to begin the argument.

To give a radically simplified version of his schema, Roman

strategy went through three stages. In the early years a continually expanding empire operated with mobile armies and a system of client states, which acted as buffers along the imperial fringe. During the second stage, which lasted for about two centuries, the client states became an integral part of the empire; Roman legions could be deployed almost exclusively along a static defensive perimeter because the interior was mostly at peace. This has sometimes been called the "eggshell strategy" — hard on the outside, soft on the inside. Then, in the third century A.D., came the disruptions caused by decade upon decade of civil war. There were about twenty-five emperors from 211 to 284; only two died natural deaths, and even those deaths would not be called happy ones. (Claudius II succumbed to the plague. The emperor Valerian led an ill-fated expedition into Mesopotamia, and was captured and used as a living footstool by the Persian ruler Shapur, who made Valerian kneel so that he could step on his back when mounting his horse.) This "sanguinary turmoil," as Luttwak calls it, proved catastrophic for Rome's defenses, and occurred at a time when for many reasons outside pressures from *barbaricum*, the surrounding barbarian world, were on the rise anyway.

Thus stage three: defense in depth. Rather than attempt to keep the whole frontier membrane intact, Rome accepted penetration as inevitable and simply sought to cope with it. Strategic points in the interior were fortified. Cities everywhere began walling themselves in; Rome's Aurelian Walls, the ones you see now, date to this time. The legions, no longer a reliable deterrent, became powerful roving armies designed to meet individual incursions as they arose. Defense in depth meant acknowledging that some parts of the empire would become battlefields, and that some parts might even have to be amputated. Defense in depth was not preventive but reactive, like antibodies; when it was triggered, the barbarians were already inside.

The drain on Roman coffers of military spending was immense. "It is impossible that our present revenues should suffice for the support of the troops, not to speak of other expenses," wrote Cassius Dio, who put this "swords or bread?" observation into the mouth of Agrippa, during the early empire; he might just as well have been speechwriting for any modern American politician wrestling with the balance between guns and butter. Luttwak observes of Rome, "The provision of security became an increasingly heavy charge on society, a charge unevenly distributed, which could enrich the wealthy and ruin the poor." He goes on: "The machinery of empire now became increasingly self-serving, with its tax collectors, administrators, and soldiers of much greater use to one another than to society at large." If you leave out the checks that Washington automatically writes to fulfill programs like Social Security and Medicare, running the military is the most expensive single thing the American government does; and as was the case in Rome, it is easily the most labor-intensive and complicated. Even as the Cold War was winding down, in the early 1990s, some three quarters of the federal government's research money was consumed by the Pentagon. Half of America's scientists and engineers were employed by military-related companies. The needs of the Pentagon skew entire industries. They cause communities to arise out of nowhere, and as suddenly to collapse. Economists argue endlessly, and inconclusively, about the impact of military spending on the American economy as a whole. Is it a depressant? A stimulant? A hallucinogen? The countless variables and their interactions defy precise sorting. The effect of military spending on government budgets is plain enough, though: investments of one kind diminish investments of another. "Every gun that is made, every warship launched, every rocket fired, signifies in the final sense a theft from those who hunger and are not fed, those who are cold and not clothed." That's not George McGovern whining in 1972. It's Dwight D. Eisenhower just stating the facts in 1950.

The United States military divides the entire world into five segments, or area commands, and it maintains basing rights (and more than seven hundred bases) in about sixty countries. In a typical year the Pentagon may conduct operations of some kind in 170 nations — "operations" that run the gamut from a handful of advisers doing drug interdiction in northern Thailand to 150,000 men and women fighting a full-fledged war on the Tigris and the Euphrates. The military's direct payroll includes almost three million active-duty soldiers and members of the reserves.

America's European Command incorporates all of the Roman Empire except Egypt and the Middle East. In school I was entranced by those sequences of maps showing the origins and growth of the Roman state from its smudge of a beginning on the plains of Latium. The last of them always depicted the empire "at its greatest extent." Mussolini affixed a four-panel marble version of this sequence to the walls of the Basilica of Maxentius, in Rome. (There was once a fifth panel, showing Mussolini's own Italian "empire," which comprised Italy, Albania, Libya, Ethiopia, Eritrea, and Somaliland; it was taken down after the war.) Rome at its greatest extent, in the second century A.D., had a perimeter of about 6,000 miles. It took in Britain south of Hadrian's Wall, followed the course of the Rhine, jumped over to the Danube, took a detour to absorb what is modern Romania, encompassed Asia Minor to the banks of the Euphrates, ran along the edge of the Sahara across North Africa, and used the Atlantic as a boundary all the way back to Britain. Little of this territory needed continual defending, any more than Kansas or Kentucky does. But the Roman military establishment left a noticeable footprint. Some three hundred military sites, mostly forts, have been identified in Britain alone.

During the civil war after the assassination of Julius Caesar, in 44 B.C., the warring commanders went into battle against one another leading armies of up to 150,000 men, a size not reached again until the Napoleonic Wars. The drain on manpower was unsustainable — one reason Augustus took advantage of the "peace dividend" to downsize the army. Augustus set the number of Roman legions at twenty-eight, and though legions might be added or subtracted, consolidated or reorganized, and supplemented with auxiliaries, this would remain roughly the size of the military for centuries, until the numbers jumped upward dramatically to deal with crises on the frontier. In practice Rome must have had something similar to U.S. military commands: Tacitus writes that at one point during the reign of Tiberius, early in the first century A.D., there were eight legions on the Rhine, two in Dalmatia, four on the Danube, three in Spain, two in North Africa, two in Egypt, and four in Syria. (That makes twenty-five; the missing three had recently been wiped out in a catastrophic encounter with a tribe called the Cherusci, in Germany.) It is customary now for military units to bear honorific names. The U.S. 4th Armored Division is known as the "Iron Horse." The 3rd Infantry Division is the "Rock of the Marne." The Roman legions anticipated this practice. There was XII Fulminata, "Thunderbolt," which served with Caesar in Gaul; XX Valeria Victrix, "Valiant and Victorious," which put down the Boudican rebels in Britain; XXI Rapax, "Grasping," like a bird of prey, formed by Octavian. The names scan like poetry — Gemina, Apollinaris, Claudia, Augusta — marching down the centuries toward an oblivion they will never reach.

When you consider the expanse of the territory to be defended, and the poor communications and sluggish pace of travel, Rome's military doesn't seem extravagant. In fact, though, it eventually had to confront a problem with a very American feel to it. "Given the worldwide array of military liabilities which the United States has assumed since 1945," the historian Paul Kennedy has written, "its capacity to carry those burdens is obviously less than it was several decades ago" — less than it was when its economic prowess was unchallenged. The Roman military was caught in the very same vise: too large to be affordable and too small to do everything it was asked to do.

A Roman Culture War — and Ours

AT SOME MOMENT in the mid to late fourth century A.D. a concerned Roman citizen sat down and began to write a long letter to his emperor. We do not know the man's name, and can only guess at the year from various internal clues. The document bears the title *De Rebus Bellicis* (*A Tract on Defense*), and it seems to have found its way to the imperial precincts and into the hands of a courtier, where it survived by a lucky fluke.

What we know about Anonymous, the author, we have to infer from the writing itself. He is educated in Latin, knows some Greek, displays a wonkish familiarity with government and policy jargon, and probably isn't a Christian. He is deeply troubled by the condition of the empire, which he considers to be in a state of material and spiritual decline. In this he is hardly alone: the late empire has given rise to plenty of such commentary. Observers wring their hands about the growth of inequality, about the way taxes fall heavily on the poor even as the rich escape, about poor administration and widespread corruption. In this instance Anonymous is writing about national security in its broadest sense, and he makes the very modern point that security isn't just a matter of raw military power but also derives from a society's overall health.

He's also worried about something very specific: the empire does not have enough people to do the work that needs to be done. In the fourth century labor shortages are a problem throughout the imperial economy — laws are passed to bind families to essential jobs, such as farming, mining, weaving, and weapons making, generation after generation — but nowhere more than in the military, which is now bigger than ever, and for which recruits are harder to get even as threats from the outside are obviously growing. What's the solution? Like some strategists in our own time — those who have invested hope in air power as a partial substitute for ground forces; those who propose walled and wired borders as a response to immigration — Anonymous seeks a technological fix. Most of *De Rebus Bellicis* is given over to a description of the new weapons of war that he is promoting. There is a warship powered not by sail or oarsmen but by oxen turning paddle wheels, which propel the ship with "a wondrous and ingenious effect." There is a chariot with whirling scythes on its axles, like the one used by Massala in the chariot race in *Ben-Hur*. There are several new types of artillery pieces, or *ballistae*.

De Rebus Bellicis has long attracted special notice because it's a rare example of Roman technological precocity. For all their applied engineering skills in some areas, especially military ones, the Romans came up short in others, and have little place in technology's annals under the heading "original ideas." "The Roman empire was technologically as backward as medieval Europe, and in some important aspects more so," writes the historian A.H.M. Jones. "Pottery was turned on the wheel, metalwork hammered out on the anvil." Measuring distances in open country was done in primitive (but surprisingly accurate) fashion: to keep track of how far they had traveled on the march, Roman armies relied on men whose job it was to walk in equal paces and count the number of steps they had taken. The technology of waterpower was understood, but more widely used for sawing marble than for industrial processes that might sustain an economy or relieve a labor crunch. It's as if America used electricity to power Disney World but not U.S. Steel. A famous study from many decades ago, by the historian Lynn White Jr., suggests that the barbarians — "far too little understood" — would have garnered a lot more patents than the Romans, maybe because they were so much less reliant on slaves to do their work for them. (And in fact the Germans today are number 3, after the Americans and the Japanese, in the number of U.S. patents granted annually.) It was the barbarians, not the Romans, who came up with such things as the making of barrels and tubs; the cultivation of rye, oats, and hops; the heavy plow; the ski; trousers; and the stirrup, which revolutionized warfare on horseback. So it is understandable that the inventions of Anonymous, colorfully rendered in thumbnail illuminations, have captured attention. Whether any of the devices would actually have worked as advertised is another matter. But they wouldn't have solved the larger military-manpower problem, any more than technology today can solve America's.

Rome's situation was acute even before the Battle of Adrianople, in 378 A.D., where a professional Roman army of 20,000 was annihilated by the Goths in the course of one hot August afternoon. When the battle was over, the emperor Valens and two thirds of his men lay dead on the field. In its aftermath the Romans had no choice but to reconstitute their lost legions the only way they could. Barbarians had for centuries been assimilated into the military in large but digestible numbers. Now, for the first time, they were invited en masse to fight under the imperial banners — and allowed to stay intact as peoples, and to occupy territory, and to be led by their own leaders. These *foederati* would turn out to be a devil's bargain. Yes, the barbarian forces often proved their mettle. At the same time, they were prone to independence and loyal to their own commanders. Their level of training did not match that of the Romans. They behaved according to different cultural standards. The chronicler Ammianus Marcellinus tells the story of one barbarian recruit who shocked his peers by drinking the blood of an enemy he had just killed. Not surprisingly, some barbarian units became free agents. The Visigothic leader Alaric rose to the rank of general in the service of Rome, fighting under the emperor Theodosius. Only when his further advancement in the hierarchy was rebuffed did he turn on the empire he had served. The historian Ramsay MacMullen writes that when Alaric and the Visigoths sacked Rome, "he and his men *were* the Roman army, and had been for decades."

Adrianople made catastrophic a set of conditions that had long been dangerous and chronic. One part of the problem was more social than quantitative. In the days of the republic the Roman military had been largely a citizen army made up of Romans themselves, with its leadership drawn from the ranks of the upper classes. That continued into imperial times, but to an ever-lessening degree. Over the years, the number of soldiers with roots in Italy declined steadily—from about 65 percent at the beginning of the first century A.D. to about 20 percent at the end. Meanwhile, the ties between the military and the civilian elite became frayed. Roman historians like Tacitus and Cassius Dio had all served in the army, but after Ammianus Marcellinus, who was alive at the time of Adrianople, the chroniclers have no firsthand knowledge of military affairs. One modern historian describes the Roman army as, increasingly, "a society rather sealed off from the ordinary, that is, from the civilian," and quotes from an ancient account of the reaction in Rome to soldiers returning from faraway frontiers: they were "most savage to look at, frightening to listen to, and boorish to talk with." America's Delta Force would fare no better in Saddle River, Brentwood, or Winnetka.

In his 1957 study *The Soldier and the State,* Samuel Huntington identified what many now describe as a growing "two cultures" problem in America, with the military class and the professional and administrative class beginning to pull apart. This was a problem that a universal military draft had kept in check, and that an all-volunteer army—meaning a paid, professional army—would only accelerate. Some 450 out of 750 Princeton graduates in the class of 1956 served in the military, including Neil Rudenstine, a future president of Harvard, and R. W. Apple, the late *New York Times* correspondent; the figure for the class of 2004 was eight out of 1,100. Huntington, himself a veteran and a member of the last generation of American scholars who did military service as a matter of routine, could as eas-

ily have written his book—*Miles et Imperium,* it would have been
called—in the reign of Diocletian as in that of Eisenhower. In
Great Britain the two-cultures divide is mitigated by the royal
family, whose young men routinely enlist in the armed forces.
But consider the Kennedy family: all four of the Kennedy brothers—Joe, John, Bobby, and Ted—served in the military; not one
of the thirty Kennedy cousins has. I haven't served in the armed
forces either, though a brother and a sister are veterans: very unusual for a middle-class family from Greenwich, Connecticut, in
the late 1970s, and nearly unheard of today.

This phenomenon has consequences for everything from military recruiting to national leadership. At a recent convention of
recruiters a distinguished scholar of manpower issues asked, "If
you had a choice between tripling your advertising budget and
having Jenna Bush join the Army, which would you take?" The
overwhelming choice was Jenna. In terms of leadership, you
can't help wondering if a late-Roman model is starting to reas-
sert itself. As a group, the most capable, well-rounded, and experienced public executives in America today are its senior military officers, not its Washington politicians. To be an American
general in effect means that you know how to run a business,
have earned a graduate degree in some serious academic subject, have gained diplomatic experience abroad, know something
about managing large numbers of people, are comfortable ne-
gotiating within the imperial bureaucracy, and understand the
infrastructure of managed violence. How many elected officials
have a comparable résumé?

The two-cultures divide has an even broader dimension: leadership aside, military society and the larger American society are
increasingly dissimilar. Military society is orderly and disciplined;
it grapples openly and pragmatically with tough issues of education, family life, poverty, and class. It's the one social segment of
America that has seriously confronted the issue of race. Only in
the military do white people routinely take orders from black

people. Military society is also deeply religious, and it's far more conservative politically than the rest of America, and far more likely to vote. Its members look with dismay at what they see as the physical and spiritual softness of civilian society. A colonel writes in the *Marine Corps Gazette:* "It is no longer enough for Marines to 'reflect' the society they defend. They must lead it, not politically but culturally. For it is the culture we are defending." There is something familiar about this colonel's discomfort: it's how Diocletian felt when he finally set eyes on Rome.

Out of necessity Diocletian expanded the legions substantially, but of course this made the overall manpower situation even more acute. How much the Roman army grew remains a subject of great debate. Did it double? Triple? And how much of the growth was real, as opposed to on paper? What is known for sure is that the army grew. Quotas had to be levied on cities and on big landowners. Then as now, physical impairment was one way out, and potential conscripts sometimes maimed themselves to escape military service—by cutting off their thumbs, for instance. But even that could prove unavailing. One emperor ordered the thumbless ones to be burned alive; another decided to take the mutilated conscripts anyway—and then doubled the quota on the localities they came from. But the competition for labor was intense. The great landed estates, the main source of wealth in an agrarian economy, needed workers. So did the *fabricae*. An arrangement with the barbarians—land, power, and autonomy in return for military service—seemed like a good idea at the time.

The New Barbarians

ONE HISTORIAN has observed that the adequacy of Rome's military "depended in fact on the success of Rome in avoiding having to face two major military threats to her frontiers simultaneously." Officially, U.S. military doctrine calls for the ability to fight two large conventional conflicts at once — a capability that the Pentagon acknowledges America no longer possesses. Meanwhile, the American military is hemorrhaging young officers; in 2005 fully a third of the West Point graduates eligible for retirement chose to leave the service at the first opportunity (after five years), up from only about a fifth not long ago. The armed forces are also facing severe recruiting shortfalls. The Army, the Army Reserve, and the Army National Guard failed to meet their recruiting goals in 2005. This state of affairs has been profoundly exacerbated by the war in Iraq. Virtually all Army and Marine combat units are in Iraq or Afghanistan, on their way back, or preparing to deploy, and the pace of rotation is proceeding at unprecedented speed. At the same time, virtually all Army National Guard and Marine Reserve combat units have been mobilized. What would happen if a crisis elsewhere demanded a significant level of commitment? One analyst has characterized the military as being in "a race against time": it must either recruit more or deploy less, or the result will be to "break" the armed forces.

The words of a modern military classic, Colonel T. R. Fehrenbach's *This Kind of War,* never lose their relevance: "You may fly over a land forever; you may bomb it, atomize it, pulverize it and wipe it clean of life — but if you desire to defend it, protect it, and keep it for civilization, you must do this on the ground, the way the Roman legions did, by putting your young men into the mud."

The American military has responded to its manpower challenges in the only two ways possible — choices that would have been familiar to a Valentinian or a Theodosius. One is by trying to raise more indigenous recruits. The other is by trying to find outsiders to help. With conscription off the table — unacceptable politically — finding more Americans isn't easy now and won't become easier. One obvious method is to lower the entry threshold. The Pentagon tried this during the Vietnam War, in an experiment called Project 100,000 — an attempt to "salvage" effective soldiers from the pool of so-called Category IV recruits, the ones who score lowest on the Armed Forces Qualification Test. The experiment succeeded mainly in introducing an Orwellian term into military parlance: New Standards Men. It was otherwise a robust failure. The New Standards Men were unfit for many specialties and therefore assigned to the infantry, making them far more likely than other recruits to die in combat. Performance was subpar in every way. They took longer to train. They were arrested more frequently. And once outside the military, their lives were no different from those of their peers who had never been in the service.

No one in the Pentagon wants to revisit the post-Vietnam years — a traumatic era for the American armed forces. The draft had been replaced by an all-volunteer military, morale was low, and the quality of recruits was abysmal. Training manuals sometimes had to be translated into comic-book form. The Pentagon announced last year that in order to meet its recruiting goals, it was raising the maximum age for Army and National Guard recruits from thirty-four to forty, and then raised it again to forty-two. It also hopes to double the enlistment bonus (to $40,000), and it is offering a $2,000 "Quick Ship" bonus for recruits who agree to start basic training within forty-five days of signature. Some Special Forces personnel in high demand are receiving re-enlistment bonuses of up to $150,000. To lower the bar for new enlistees even further, the military has softened its educational requirements. In 2005 the military accepted more recruits without high school diplomas than it had in years, and it also accepted more Category IV recruits. In addition, the military has eased the ban on tattoos, relaxed its drug-testing policies, and waived the obesity cutoff if a potential recruit can pass a somewhat forgiving fitness test. One recruiting officer explains, "There's just a lot of kids sitting around playing Xbox and eating junk food." Niall Ferguson writes in Colossus that for Americans, "the white man's burden is around his waist." Overall, the recruits accepted into the military in 2005 were the least qualified in a decade.

The military is also making it just plain harder to leave the

service. A Pentagon memo in 2005 ordered commanders to reduce high attrition ("We need your concerted effort to reverse
the negative trend") by paying less attention to poor performance, bad behavior, and unruly body mass. Recruiters are looking beyond the traditional American heartland. A recent newspaper account began, "From Pago Pago in American Samoa to
Yap in Micronesia, 4,000 miles to the west, Army recruiters are
scouring the Pacific, looking for high school graduates to enlist
at a time when the Iraq war is turning off many candidates in
the States." Another source that could be tapped is illegal aliens.
A bill now in Congress targets the noncitizen children of undocumented immigrants: they would get legal status and become
eligible for citizenship if they signed up for two years in the military. The military-affairs analyst Max Boot, taking a page from
ancient imperial practice — Roman citizenship after twenty-five
years in the army — has proposed conferring American citizenship in return for a period of military service: "We could model
a Freedom Legion after the French Foreign Legion. Or we could
allow foreigners to join regular units after a period of English-
language instruction, if necessary." (He went on to ask, "Anyway, what's the alternative? $100,000 signing bonuses? Recruiting
felons?")

Outsiders could come in other forms. There are existing alliances — such as, for instance, NATO. But NATO is a tempera-
mental instrument, and the forces of its member nations are already stretched thin. To wage the war in Iraq, the United States
cobbled together an ad hoc thirty-four-member Coalition of the
Willing for both military and political purposes, but even those
making significant commitments (Britain, Italy, Spain) were
quickly under pressure to bring their forces home. Others, like
Honduras, Moldova, and the Kingdom of Tonga, sent only to-
ken forces to begin with, and have completely withdrawn. The
Kingdom of Morocco, a coalition member, sent no troops but
made available 2,000 monkeys for the purpose of helping to find

and detonate land mines. The Mongolians, who actually sacked Baghdad in 1258 A.D. but had not been back since, still have some 130 soldiers in Iraq.

In every way the "outside" group that matters most to the military, by an order of magnitude, is actually its civilian work force. The true analogue of the "barbarization" of Rome's armed forces is the "civilianization" of America's. Yesterday's Conan the Barbarian is today's Conan the Contractor. Civilian contractors performing military chores number in the hundreds of thousands; America makes greater use of the so-called "privatized military industry" than any other nation. Some of these contractors supply low-end support services. Security at the U.S. Military Academy, at West Point, is provided by Alutiiq Security and Technology, a subsidiary of Wackenhut. (A guard at the gate told a recent visitor, "If you don't want to take the physical you still get $20 an hour to stand here overnight. I figure, if a bomb goes off the $20 wasn't worth it, but otherwise it's a good deal.") Other contractors could just as readily field entire armies. America's military contract work force is just one component of an international industry: companies like MPRI, Airscan, Dyncorp, and Kellogg Brown & Root can variously deploy troops, build and run military bases, train guerrilla forces, conduct air surveillance, mount coups, stave off coups, and put back together the countries that wars have just destroyed (or at least try to). "Frankly, I'd like to see the government get out of the business of war altogether and leave the whole field to private industry"—Milo Minderbinder's remark in *Catch-22* was meant to be absurd, but it actually describes a military system that is coming into being.

The Iraq War is the most privatized major conflict since the Renaissance: many tens of billions of dollars have already gone to private contractors involved in the invasion and the occupation. Centuries ago it was not unusual for monarchs and city-states to contract out much of their war-fighting business to professional mercenaries; the term *condottieri*, referring to men of this kind, comes from the Italian word for "contract." The Swiss Guards standing decoratively with their halberds at the Vatican had their origin as mercenaries, which is what they still are. But contractors declined in importance with the rise of large standing armies. Now, in America, their influence is growing, fostered by recruitment problems and by the limits the Pentagon feels it must maintain, for political reasons, on the official size of the armed forces overall. On the books, some 150,000 U.S. military personnel are involved in the occupation of Iraq at any moment; that number excludes another 100,000 men and women from the private sector, hired directly or indirectly by the Pentagon or other government departments.

They may be American . . . or British or Egyptian or South African. They advise on safety at Iraqi police recruiting stations. They operate prisons. They guard munitions depots. They provide security for individuals (including some American generals), for supply convoys, and for entire military bases, as well as for all the other contractors trying to build power plants and hospitals and schools. They are a highly visible presence in Iraq, speeding around in armored SUVs. Some of the security companies have an established pedigree. Others, like Triple Canopy, based in Chicago, which was founded by several Special Forces veterans soon after 9/11, are aggressive young start-ups. The first armored cars Triple Canopy deployed were Mercedes-Benz sedans whose previous users had been an assortment of rap stars and the sultan of Brunei.

Roman commentators had a standard bill of complaint against the barbarian forces in the military: Their training did not match that of the Romans. There were important issues of reliability and trust. They were hard to discipline. And inevitably, when Roman legions saw what the barbarian auxiliaries could get away with, their own standards began to slip. Training in full armor became a thing of the past, and the consequences for "readiness" cascaded downward from there. In the late fourth century the Roman military writer Vegetius described the real-world calamities that bad habits produced:

> But when, because of negligence and laziness, parade ground drills were abandoned, the customary armor began to seem heavy since the soldiers rarely ever wore it. Therefore they first asked the emperor to set aside the breastplates and mail and then the helmets. So our soldiers fought the Goths without any protection for chest and head and were often beaten by archers. Although there were many disasters, which led to the loss of great cities, no one tried to restore breastplates and helmets to the infantry.

A recent study by the military analyst Andrew Krepinevich raises all the same concerns. Contractors operate outside military justice — they can't be disciplined. They also operate under different rules of engagement (if any). Their pay and living conditions are far better than those of ordinary soldiers — bad for morale. They can quit whenever they want, and go work for somebody else. They sometimes need and deserve access to military information, which raises concerns: "The problem here, of course, is whether the intelligence will remain a secret," Krepinevich writes. There's no disputing, though, that these contract *foederati* are here to stay. The U.S. military cannot operate without them. But their services are for sale.

In the spring of 2006 J. Cofer Black, a former counterterrorism chief at the Central Intelligence Agency and now the vice chairman of the private security firm Blackwater USA, addressed a meeting of the Special Operations Forces Exhibition and Conference, held in Amman, Jordan. Black in essence put his company at America's (or *someone's*) disposal as an army for hire. Manpower shortages will only get worse. National armies are sluggish. Their deployment is politically fraught. Why not a brigade-sized alternative, created by Blackwater and led by the former special-operations, intelligence, and law-enforcement personnel who constitute much of Blackwater's professional pool?

"It's an intriguing, good idea from a practical standpoint, because we're low-cost and fast," Black explained. He went on, sounding the plaintive note one might have heard from Alaric: "The issue is, who's going to let us play on their team?"

3

THE FIXERS

When Public Good Meets
Private Opportunity

Four or five men get together, they think up some way
to fool the emperor, and they inform him of whatever he
must approve. . . . As Diocletian himself used to declare, an
emperor — good, careful, the very best — is put up for sale.
— *Historia Augusta,* c. early fourth century A.D.

Corruption charges. Corruption? Corruption ain't nothing
more than government intrusion into market efficiencies
in the form of regulation. That's Milton Friedman. He got
a goddamn Nobel prize. . . : Corruption is our protection.
Corruption is what keeps us safe and warm. . . . Corruption is how we win.
— the character Danny Dalton, in the movie *Syriana*

T HERE IS AN OLD medical museum in Rome not far from
the Vatican, a place that Leonardo da Vinci and Martin Luther would have known, and a few years ago I was taken
there by a friend. It's up under the rafters of a fifteenth-century
ospedale, and the creaky wooden cabinets hold unfriendly im-
plements (surgical, obstetrical, juridical) going back to ancient
times. In one dark corner I noticed a small display case, and under the glass, on faded, musty velvet, lay half of a human mandi-
ble and a fragment of human cranium. My friend leaned over to

read the label, written in Italian in an elegant nineteenth-century hand. He looked up in surprise and said, "Pliny the Elder."

I'd never really made the acquaintance of Pliny the Elder — the great naturalist who set out by ship to see the eruption of Vesuvius at close range, and succumbed to the fumes. His nephew was another story: I felt as if I'd known him for years. Pliny the Younger (c. 63–113 A.D.) was born during the reign of Nero and died during the reign of Trajan, and he left behind a collection of letters, hundreds of them, written in a style that glides easily across the centuries. He is alive on the page in a way that appeals to a high school Latinist weary of Virgil's formal meter and Cicero's love for the sound of his own voice. Combine the catholic tastes of a Montaigne, the political connections of a Galbraith, and the amiable grace of a George Plimpton, and throw in some of the fussiness of Felix Unger and the lordly self-esteem of Mr. Toad — and you have the letters of Pliny the Younger. Pliny's life was enviable. He was born to wealth, and followed the sequential steps of the *cursus honorum,* the escalating series of offices a public man in Rome was expected to hold — in essence, the résumé of a George Bush the Elder up to 1980. The historians Tacitus and Suetonius were friends of his. Trajan, the emperor, was a correspondent. And in his letters Pliny writes about almost anything at all: the fate of his uncle at Vesuvius, his gardens and his villa, the careers of friends and enemies, reading poetry aloud, the death of a slave. Sent near the end of his life to be the emperor's special envoy to Bithynia, a troubled province in the East, Pliny pesters Trajan all the time (lucky for us) about the most humdrum matters — imagine if MacArthur in Tokyo had felt he must check with Truman about where to billet some visiting British admiral, or whether to let the Japanese play baseball. Trajan almost always writes back with judicious advice and patient forbearance. Some might even hear mild amusement. Occasionally he tells Pliny to get a life.

Amid all this variety of correspondence one kind of letter stands out: the patronage letter, seeking help or a job or a break for some member of Pliny's extended personal network. Many of these are addressed directly to the emperor. Pliny requests a praetorship (a public office) for a man named Attius Sura, "now that there is a vacancy"; he urges the granting of a petition to build new public baths at Prusa, because the existing ones are "old and dilapidated"; he notes the burning down of the Elder Citizens Club in Nicomedia, and asks permission to form a company of firemen; he urges a military promotion for the son of an army friend; he seeks permission for his wife, because of a death in the family, to travel using the facilities of the *cursus publicus,* the imperial post, with its system of carriages and way stations normally reserved for official business; and he intercedes for athletes in the Triumphal Games, who wonder if they can receive their prize money right after winning, instead of having to wait until the formal celebration when they return to their native towns.

Here's a letter to the emperor on behalf of a "resident alien" physician who had come to Pliny's aid:

> When I was seriously ill last year, Sir, and in some danger for my life, I called in a medical therapist whose care and attentiveness I cannot adequately reward without the help of your kind interest in the man. I pray you therefore to grant him Roman citizenship.

Here is another letter to Trajan, seeking to confirm a promise that had been made but was left unfulfilled by the previous emperor, Nerva:

> Your kindness, noble emperor, of which I have full personal experience, encourages me to venture to ask you to extend it to my friends, among whom Voconius Romanus has the highest claim. He has been my friend from our early years when we were at school together, and for this reason I petitioned your deified father, the late Emperor, to raise him to the dignity of senatorial rank.

Pliny was hardly alone in such exertions. A notable proportion of the surviving correspondence of any prominent Roman consists of letters of recommendation and letters that bestow favors or ask for favors in return. The very term "letters of recommendation" — *litterae commendaticiae* — comes from Cicero, who in one of his works even pulls together a collection of his own for use by others as models. (Given the creeping fulsomeness that tends to infect such letters, an all too familiar arms race of half-truth and embellishment, Cicero also used a secret code with one of his correspondents to indicate when he was just doing his duty and when he really meant it.) All these kinds of letters exemplify the networks of patronage that held Roman society together, from top to bottom, from the margins to the core, starting in the city-state's very earliest days and running through the empire's very last and beyond. The wheeling and dealing, the back-scratching, the logrolling, the influence peddling — at every level of the Roman social scale they're at once intense and routine.

Meet Marcus Cornelius Fronto, orator and advocate, grammarian and flatterer, a native of North Africa who achieved renown and wealth in Rome and became *magister imperatorum*, a teacher of the future emperors Marcus Aurelius and Lucius Verus. He was the consummate courtier, a florid writer, and something of a pedant, who loved dredging up obscure words in a William F. Buckley–esque way. His correspondents included not only Marcus and Lucius but also their adoptive father, Antoninus Pius, the reigning emperor. That his letters (along with some replies) survive at all is a happy accident. They were collected into a codex in late antiquity, but the pages were torn from the book and reused by monks, who wrote over them to record the acts of the Council of Chalcedon, in 451. These palimpsest pages weren't discovered until the nineteenth century. Fronto's letters offer glimpses of daily life in the imperial palace — Marcus Aurelius playing with his toddlers, for instance: "I have seen your little chicks, and a more welcome sight I shall never in my life see, so like in features to you that nothing can be more like than the likeness" (a charming vignette, as the editor of Fronto's letters points out, until you remember that one of the little chicks will turn into that vile rooster, the emperor Commodus). Fronto comes across by and large as a warm-hearted fellow. And we frequently see him in the role of *patronus*: taking the children of provincial friends into his home; introducing them around and advancing their careers; and interceding with judges to vouch for the character of clients whose cases are pending.

Usually he writes with a literary flourish, as in a letter to an influential friend on behalf of a young man named Volumnius:

> If you love me, accord to Volumnius so much respect and opportunity of gaining your friendship, for very dear friends have enlisted my sympathy for him. Therefore I would ask you to welcome him with such kindly friendship as the great Achilles wished to show, when he bid the son of Menoetius *mix the wine stronger.*

And here, two centuries later, is the rhetorician Libanius, a native of Antioch and a supremely cultured sophist: friend of the emperor Julian, mentor of the historian Ammianus Marcellinus, author of a self-absorbed and sometimes self-pitying autobiography, and a man intensely loyal to a wide circle of associates. Some 1,500 of his letters survive, many of them interventions in his role as patron. Here he asks a powerful governor to come to the aid of a man Libanius hardly knows, but who is a member of the family of his esteemed teacher Zenobius.

> The bearer, his relative and namesake, was appointed justice of the peace, and although he has taken especial care of the city, he has been attacked and ejected from his post . . . My request is that the unjustified usurper should himself be justifiably ejected, so that I may do my duty by my dead teacher.

The relationships built through the Roman system of patronage were the basic foundations of power. Like a lattice they formed a support structure for all of society, broad and hierarchical. Patronage linked the members of the tiny elite to one another, and the large mass of the powerless to people at the apex. Over time, however, the system would go horribly wrong, its central strength — that all-pervading nature — becoming its most dangerous characteristic. The change came when money entered the picture in a significant way, and these countless transactions began to come at a price. Libanius, a gifted manipulator of the patronage system, would in one of his letters cynically liken the men around the emperor to "that cloud of Zeus from which he rained down gold." Tainted by venality, patronage mutated from an instrument of cohesion into one of corrosion.

The slow evolutionary spiral can be seen in the history of a single Latin term. A little more than fifty years ago the Oxford historian Geoffrey de Ste. Croix, a radical thinker and formidable classicist, decided to take a close look at the change in connotation over five centuries of the Latin word *suffragium*, which originally meant "voting tablet" or "ballot." That change, he concluded, illustrated a change in something fundamental about the character of Roman society and its "inner political evolution."

What was the change? The original meaning went back to the days of the Roman Republic, which had possessed modest elements of actual democracy. The citizens of Rome, by means of the *suffragium*, could exercise their influence on some kinds of decisions, such as electing people to certain offices. In practice, the great men of Rome controlled large blocs of votes, corresponding to their patronage networks. Over time Rome's republican forms of government calcified into empty ritual or withered away entirely. *Suffragium* meaning "ballot" no longer served any real political function. But the web of patrons and clients was still the Roman system's substructure, and in this context *suffragium* came to mean the pressure that could be exerted on one's behalf by a powerful man, whether to obtain a job or to

influence a court case or to secure a contract. To ask a patron
for this form of intervention and to exert *suffragium* on behalf of
a client would have been a routine social interaction, and in its
pure form no money would be expected to change hands. This is
the relationship you see in the letters from Pliny and the others.
(You see it, too, when Christianity begins to take on the characteristics of its Roman environment: *suffragium* comes to be the
word used when believers, seeking salvation after death, describe
how they ask the saints to intercede in their behalf.)

Now stir large amounts of money into this system. It is not
a great conceptual distance, Ste. Croix observes, to move from
the idea of exercising *suffragium* because of an age-old sense of
reciprocal duty to that of exercising it because doing so could
be lucrative. And this, indeed, is where the future lies, the idea
of *quid pro quo* eventually becoming so accepted and ingrained
that emperors stop trying to halt the practice and instead seek to
contain it by codifying it. Thus, in the fourth century, decrees are
promulgated to ensure that the person seeking the *quid* actually
delivers the *quo*. Before long, *suffragium* has changed its meaning
once again. Now it refers not to the influence brought to bear
but to the money being paid for it: "a gift, payment, or bribe."
By empire's end, all public transactions require the payment of
money, and the pursuit of money and personal advancement has
become the purpose of all public jobs, which of course themselves cost money to acquire.

Looking back at the story of *suffragium,* from ballot box to
cash box, Ste. Croix composes this epitaph: "Here, in miniature,
is the political history of Rome."

What's in It for Me?

IN TRUTH, THE ARC traced by *suffragium* is a carapace covering
not just the political history of Rome but its social, military, and
every other kind of history. It goes to the heart of a question that
is starting to be asked in America, where some form of degen-

erative neuropolitical condition has left government responsive
to particular interests but deaf to the popular will. The question
is this: Where is the boundary between public good and private
advantage, between "ours" and "mine"? From this question others follow. What happens when public and private interests are
not aligned? Which outsiders, if any, should be allowed to put
their hands on the machinery of government? How can governments exert collective power if the levers and winches and cogs
lie increasingly outside public control?

The phenomenon with which all these questions intersect
was called the "privatization of power," or sometimes just "privatization," by Ramsay MacMullen in his important study *Corruption and the Decline of Rome*. MacMullen's subject is "the diverting
of governmental force, its *mis*direction." In other words, how
does it come about that the word and writ of a powerful central government lose all vector and force — like a car in which
the steering wheel and accelerator pedal are suddenly connected
to nothing at all? Serious challenges to any society can come
from outside forces — environmental catastrophe, foreign invasion. Privatization is fundamentally an internal factor, though it
has an impact on the ability to face external threats. MacMullen
asked his question — How does power become powerless? — out
of dissatisfaction with the many theories put forward to explain
Rome's gradual decline in the West. His answer is privatization — the deflection of public purpose by private interest. Such
deflection of purpose occurs in any number of ways. It occurs
whenever official positions are bought and sold. It occurs when
people must pay before officials will act, and it occurs if payment
also determines *how* they will act. And it can occur anytime public tasks (the collecting of taxes, the quartering of troops, the
management of projects) are lodged in private hands, no matter
how honest the intention or efficient the arrangement, because
private and public interests tend to diverge over time. Privatization, whether legal or corrupt, is how the gears of government

come to break. In Rome, the consequences were felt in every area of society.

Let's start with how the Roman system worked during the many centuries when it actually did. By modern standards there were not a great many officials or bureaucrats in Rome until late in the empire; the administration and well-being of the capital and all the other cities and towns depended on the talents and the largesse of the upper classes, and on the patronage networks they controlled. In the system's idealized form, the elites and their clients constituted an interlocking force for cohesion. A memorable passage in Jerome Carcopino's *Daily Life in Ancient Rome* describes what happened every morning soon after Romans woke up, when all around the city clients visited their patrons, and each was alert to the other's needs. Does that famous early scene in *The Godfather* — the don's audience with supplicants in his shuttered study as the wedding reception unfolds outside — capture some of this flavor? On those very rare mornings when I've found myself sipping $15 orange juice at the Four Seasons in Washington or New York, I've enjoyed imagining the breakfast convergences at tables all around me as an elite remnant of the old Roman dynamic. But to get Rome right you'd have to extend the scene to every suburban Hyatt, every neighborhood diner; you'd have to see these relationships governing every business transaction, every trip to the doctor's office, every college application, not to mention the selection of your spouse, the fate of your litigation, the delivery of your newspaper. Each person had a place in the chain somewhere, and most people had more than one. Carcopino writes:

> From the parasite do-nothing up to the great aristocrat there was no man in Rome who did not feel himself bound to someone more powerful above him by the same obligations of respect, or, to use the technical term, the same *obsequium*, that bound the ex-slave to the master who had manumitted

him. The *patronus,* for his part, was in honour bound to welcome his client to his house, to invite him from time to time to his table, to come to his assistance, and to make him gifts.

Patrons and clients were stock targets of comedy and satire. The patron-client relationship was so pervasive that it helps illuminate not only Rome's social architecture but also, frequently, its way of conducting foreign affairs. The term "client state" came into being for a reason. As Julius Caesar fought his way through Gaul, he brought tribal chieftains over to his side, and described their professions of loyalty to him—and thus to Rome—as those of clients to a patron. The relationships of the Bush family with various world leaders have often been essentially personal. The longtime Saudi ambassador to Washington, Prince Bandar bin Sultan bin Abdulaziz al-Saud, spent so much time at Bush family gatherings that he came to be known as "Bandar Bush."

Patronage spilled over into communal adornment; it was in fact inseparable from it. The Roman magnates competed with one another to endow the capital with improvements. Rome was a city with many wealthy patrons—indeed, many absurdly wealthy patrons. The class stratification of Roman society was extreme; by comparison, Victorian England might seem a laboratory of equality. Rome's wealthiest class, the senatorial aristocracy, constituted by one estimate two thousandths of one percent of the population; then came the equestrian class, with perhaps a tenth of a percent. Collectively these people owned almost everything. Americans are well aware of the nation's worsening income inequality, with those in the top one percent earning fifty times more a year than those in the bottom 20 percent. The latest data show the differential getting wider. But in the Roman Empire the gap between the few thousands who constituted the upper elite and everyone else was on the order of 5,000 or 10,000 to one. In the city of Rome the rich occupied sprawling estates, the sumptuous *horti,* which ringed the dense urban core — garden districts, some of them, to this day. The expectation in Rome, maintained over many centuries, was that affluent citizens, as individuals rather than as taxpayers, should provide for community needs. Did the city require another aqueduct? An emergency supply of wheat? A fountain? New roads? Baths? A stadium? A temple? Repairs to the walls? Some magnate would surely provide it — in return, implicitly, for a measure of public power, and of course for ample public recognition. Inscriptions on countless marble fragments attest to such generosity — an early version of "brought to you by . . ." You can't sit drinking an espresso in front of the Pantheon without noticing that you have M. Agrippa (the name rendered in very big letters) to thank for the original building. The historian Cassius Dio describes how this same Agrippa in a single year repaired many of the public buildings in Rome, put statues in the public baths, distributed salt and olive oil to the masses, paid barbers to give everyone haircuts, and after cleaning out Rome's great sewers sailed a small boat through their main channel, the Cloaca Maxima, and out into the Tiber.

The private responsibilities of the wealthy extended to public spectacles in the arena. Take someone like Quintus Aurelius Symmachus, the prefect of Rome in the late fourth century, a man of real substance and vast resources (he owned three houses in Rome and twenty-eight houses elsewhere in Italy), and with an elevated, even oxygen-starved, sense of his own status (he once referred to the Roman aristocracy as "the better part of the human race"). Closer to our own era, Symmachus might have been an Averell Harriman or a Nelson Rockefeller. And yet you find him fretting in one of his letters about whether the African beasts the emperor has promised will arrive in time. Elsewhere he complains about the customs duties he must pay on imported bears, and about the problem posed by some crocodiles that are refusing to eat. In one letter Symmachus notes that twenty-nine Saxon captives being sent to him had strangled one another rather than face combat and death in the gladiatorial games, and goes on to wonder if perhaps this wasn't a blessing: the slaughter of all twenty-nine at once might have pushed the boundaries of good taste — might have been, to use his word, a bit "gaudy."

On Rome's edifice of private giving — whether with the seemliness of an Andrew Carnegie or the vulgarity of a Donald Trump — an empire was built. An ethic of philanthropy and public service spread from city to city throughout the empire. It was upheld in Rome by the senatorial aristocracy, and elsewhere mainly by the type of citizen who constituted the curial class — a group that included major property owners and professionals. These were people who tried to live upstanding lives according to the mores of the day, who controlled the bulk of the local wealth, and who assumed the burden of public office. Think of them, perhaps, as something like the provincial aristocracy in John O'Hara's short stories and novels — the upper stratum of O'Hara's fictional Gibbsville, evoked in one recent essay as follows: "A few families control the economic, political, social, and cultural life of their localities. Their sons may attend nearby prep schools and colleges, or they may go off to, say, Lawrenceville and Princeton, but most return to run the real-estate concerns, banks, law firms, factories, and mills that are at the center of the economy and from which their families derive power and prestige."

The Roman system was a remarkable contrivance. But it contained the seeds of its own destruction. For one thing, it fostered an expectation that "others" would always provide. If public amenities came into being through private munificence — and if these in turn served to enhance private glory — then why should the public pay for their upkeep? This way of doing business "did not work for the common benefit of the overall urban fabric," writes one historian, much less nurture a sense of common purpose and shared obligation. I've seen the same mindset at work within my own state, Massachusetts, in hardscrabble mill towns whose philanthropic founding families have departed, where local taxpayers resist the idea that support of libraries and hospitals must now rest with the community as a whole. Moreover, even at its most uncorrupted, the patronage system was greased by small considerations: "It was a genial, oily, present-giving world," Ramsay MacMullen observes. Pliny writes to a friend: "Thank you for those fieldfares, but being at Laurentum I can't match them with anything from town, nor can I send you any fish as long as the weather is bad." A man named Gemellus sends instructions to his son about how to coddle a certain official: "Send him an artaba of olives and some fish, as we want to make use of him."

Now gradually remove from all this any sense of public spirit or public obligation and replace it at every level of government—in the barracks, the courts, the city councils, the provincial prefectures—with an attitude of "What's in it for me?" To see this transition in starkly American terms, first consider the idealistic sensibility—wise friends promoting wise friends in a noble venture—of a letter of recommendation written by Benjamin Franklin to George Washington in 1777, on a matter of public business:

> Sir,
>
> The Gentleman who will have the Honour of waiting upon you with this Letter is the Baron de Steuben...He goes to America with a true Zeal for our Cause, and a View of engaging in it and rendring it all the Service in his Power. He is recommended to us by two of the best Judges of military Merit in this Country.

For comparison, consider the more contemporary sentiments in e-mails written by a Washington lobbyist, also on a matter of public business: in this instance, mounting a political operation to reopen the Speaking Rock Casino, in Texas, in return for millions of dollars in fees and political contributions. The lobbyist, Jack Abramoff, explained to his clients, the Tigua Indian tribe:

> This political operation will result in a Majority of both federal chambers either becoming close friends of the tribe or fearing the tribe in a very short period of time. Simply put you need 218 friends in the U.S. House and 51 Senators on your side very quickly, and we will do that through both love and fear.

Of course from time to time Zeal for the Cause might need to be topped off:

> Our friend . . . asked if you could help (as in cover) a Scotland golf trip for him and some staff (his committee chief of staff) for August. The trip will be quite expensive (we did this for another member — you know who 2 years ago). Let me know if you guys could do $50 K.

This is the story MacMullen traces, as throughout the empire a lubricious glaze of venality came to coat every governmental surface. I don't know how it would be phrased in Latin, but one of Jack Abramoff's e-mails ("Da man! You iz da man! Do you hear me?! You da man!! How much $$ coming tomorrow? Did we get some more $$ in?") captures some of the spirit of public service in the late empire. What accounts for the change? No one factor, MacMullen believes, but some combination of many: the sheer growth in the government's administrative reach; as a result, the transformation of "public service" from the rotating duty of the curial class into a lifelong career for a larger group; the flight of the elite from public service anyway, because the demands could prove so onerous; the ambiguity of many laws, allowing money to sway judgments; the increasing severity of punishments, which people would pay anything to escape; and the generally poor communications, which among local officials abetted a sense of impunity ("What happens in Bithynia stays

in Bithynia"). A bronze plaque was affixed to a public building
in Timgad, in Numidia, a city built as a bastion against the Berbers, which literally provided a recommended price list for payments to ensure the prosecution and success of various kinds of
litigation. We don't have anything quite like that now, I suppose,
but have you ever received a fundraising solicitation from one of
the political parties, with degrees of access and other perquisites
tied to specific contribution levels? Here's the Republican contribution hierarchy for the 2004 elections, which I can't help visual-
izing as a bronze plaque:

$300,000	SUPER RANGER
$250,000	REPUBLICAN REGENT
$200,000	RANGER
$100,000	PIONEER

Time and again imperial decrees throughout the later empire attempt to put a stop to skimming, extortion, and the illicit use of
office; or, failing that, to codify what may be permissible. But
the emperors are standing athwart the tide, and the imperial
pronouncements have a doomed, forlorn, ritual feel to them.
Modern newspaper headlines like "Congress Votes New Curbs
on Lobbyists" convey something of the same formulaic quality.

How does the buying and selling of influence hollow out
government? Some make the argument that whatever its moral
shortcomings, the profit motive, including its corrupt dimension,
is in fact an efficient economic mechanism—it's nothing if not
results-oriented. It gets things done. It's *how we win*. But as Mac-
Mullen points out, for a government to be systemically effective
on a national or an imperial scale, there needs to be a presumption that information is traveling accurately up and down the administrative chain of command, and that every link in the chain
between a command and its execution is reliable and strong. "Orders have to be followed throughout the whole train of power
that originates in the imperial palace and reaches, at the end, to
a hundred cobblers in the Bay-of-Naples area, a hundred peasant

owners of ox-carts in Cappadocia," MacMullen writes. "At every point of connection the original intent must be transmitted as it was received. Otherwise it will come to nothing."

Putting power into private hands — with or without some financial consideration — frequently ends up breaking that link. Making the exercise of power contingent on payment breaks the link by definition. In Rome's case, one of the functions that suffered most acutely from corruption was national security. Consider the story of the citizens of Lepcis, in what is now Libya, and their treatment at the hands of a man named Count Romanus, the military commander of northern Africa. The episode, cited by MacMullen, is recounted at length by both Ammianus Marcellinus and Edward Gibbon. It demonstrates how seemingly self-contained instances of venality can have unpredictable ramifications that do enormous harm.

In the year 363 A.D. the residents of Lepcis — one of a cluster of three cities in northern Libya that gave rise to the ancient collective name Tripoli, which of course we still use — found themselves under assault by Berber tribesmen, who ravaged the countryside and took away hostages. The Lepcitanians looked to Count Romanus for relief; he was, after all, the emperor's man on the scene. "His abilities were not inadequate to his station," Gibbon remarks of the count, but unfortunately, "sordid interest was the sole motive of his conduct." He regarded his government office as a license to pursue personal objectives, and so when he arrived at Lepcis he demanded payment — 4,000 camels, to begin with — before he would consent to hunt down the Berbers. Lepcis refused, Romanus withdrew, and the attacks continued.

The people of Lepcis were not disposed to tolerate this affront. Lepcis was a rich and significant metropolis — the birthplace of the emperor Septimius Severus. If Libya ever succeeds in jump-starting a tourist industry, it will owe much to this city's magnificent seaside remains. The citizens immediately sent deputies to the emperor Valentinian, in Trier; perhaps the audience was held in the great imperial basilica, built by Constantine, that still stands in the city center. The emperor decided to dispatch a trusted aide named Palladius to Africa to investigate; and with Palladius he also sent a large sum of money to pay the army's wages. But, as Gibbon writes, "the rigid impartiality of Palladius was easily disarmed." Upon arrival in Lepcis, he decided to keep some of the money for himself. Count Romanus found out, and used this as leverage to influence the report Palladius carried back to the emperor. The count was exonerated, and several of his accusers — upright and truthful men — were executed. Two others had their tongues cut out.

Because Romanus had escaped justice, he remained in good standing as the military commander of Africa. Ten years went by, and during that time his extortionary actions fed a revolt by a Moorish leader and client prince named Firmus. The revolt was so serious that the emperor had to send Theodosius, the army's *magister militum,* or chief of staff, to Africa to put it down. To obtain the legions to accomplish this, Theodosius borrowed from the Roman forces on the Danube, leaving that critical frontier weakened and exposed. Not surprisingly, barbarian tribes exploited the opportunity and invaded Roman territory.

In Africa, Theodosius removed Count Romanus from command and quashed the rebellion. He also made a discovery: among the count's papers Theodosius found documents showing that Palladius had lied to the emperor ten years earlier. Palladius was arrested in Italy, and sensibly committed suicide while on his way to face Valentinian in Trier.

Looking back, let's consider the calculus of corruption: the actions of one greedy soldier and one greedy bureaucrat together caused serious breaches of public order on two frontiers. Now imagine this way of doing business extended empire-wide. Imagine it encompassing everyone, from farmers and craftsmen up through local magistrates and imperial governors. Imagine

it going on for a hundred or two hundred years. Thus, for instance, by the fifth century A.D. unit commanders are routinely
sending inflated rosters to headquarters for payment on the basis of head count, keeping the difference for themselves. Soldiers
are allowed to go away on leave and then disperse into the civilian populace, but are kept on the rosters. The on-paper size of
the army is vast; the true size is smaller—and in fact unknow-
able. That army needs supplies—boots, olive oil, iron—and
from source to destination the supplies are skimmed and sold,
causing shortages down the line and sometimes mutiny. It had
once been customary, when accepting barbarians into the empire as *foederati*, to demand troops from them in return, in order
to augment the empire's dwindling stocks of manpower. Why
not just take cash instead? This approach was certainly favored
by those who would handle the money. But it made the military
no larger.

The People's Business, Inc.

IN THE END, Rome was heading toward something the Romans
couldn't, by definition, have a term for. But we do: it's the Middle
Ages. The precise definition of "feudalism" is one of those things
on which medievalists can't quite agree—the field is divided into
warring fiefdoms—but the historian F. L. Ganshof discerned in
feudal society one basic quality: "a dispersal of political authority
amongst a hierarchy of persons who exercise in their own interest powers normally attributed to the state." In other words, the
public interest had become private.

This isn't the place for an extended excursion across a thousand years of Western history. In brief, for many centuries power
was wielded in Europe by monarchs and vassals as if it were a
form of private property. The levying of taxes, the raising of
armies, the meting out of justice—these things were done in
the name of the ruler, and the fruits of his administration were
enjoyed by those who acknowledged the ruler's personal lord-

ship. The eventual path away from the Middle Ages was marked by the halting emergence of governments defined by communal interest rather than private prerogative. Power was no longer justified as simply a form of property. Social services and protections, and your rights as a person, became consequences of citizenship, not of a deal between a magnate and his underlings, or between a private entity and its clients. America came into being toward the tail end of this process; the Mayflower Compact straddles old and new, beginning with a pious nod to the king but then going on to set up a "civill Body Politick" effectively without one, which America as a whole would emphatically do a century and a half later. To most modern eyes, the general tendency described here—away from power as private property and toward power as public authority—has been seen in a positive light: evolution's arrow was pointing the way it should.

But sometime in the late twentieth century the arrow began changing direction. It began indicating a deflection of power back into private hands. Privatization today sometimes makes itself felt in ways that would have turned no heads in ancient Rome. It of course still includes influence peddling and bribery and the buying and selling of public office. Representative Randy ("Duke") Cunningham, now in jail, infamously drafted a "bribe menu" on official stationery, linking the size of defense contracts he would deliver with the size of payments he received. Representative Bob Ney, implicated in the Abramoff scandals, has left Congress, having been warned by his majority leader that if he stayed he "could not expect a lucrative career on K Street"—that is, he would jeopardize any future as a *suffragator.* And as in Rome, privatization still includes turning over government departments to incompetent cronies, empowering private individuals at the expense of public intentions. The leadership of the Federal Emergency Management Agency, which proved unable to cope with the Hurricane Katrina disaster, is only the most prominent instance.

But the dominant form of privatization today is something

relatively new, at least in its dimensions. Government on its stupendous modern scale — regulating every industry; redistributing treasure from one sector of society to another; forecasting the weather and engineering the genome; reaching bureaucratically into every intimate cranny of life — simply did not exist in ancient Rome. Because the extent of government is larger, privatization has more scope. Its most pervasive form is perfectly legal: the hiring of profit-making companies by the thousands to do government jobs, and sometimes the wholesale reflagging of government agencies into private or semiprivate companies. The ostensible motives may be pure, but the effect in every case is to insert an independent agent, with its own interests to consider and protect, into the space between public will and public outcome — a dynamic that represents a potential "diverting of government force" far more systemic and insidious than outright venality.

Privatization along these lines has occurred most decisively in America and Britain. In 1976 a book was published in the United States called *The Shadow Government;* its subtitle spoke ominously of "the government's multi-billion-dollar giveaway" of decision-making authority. In tone the book comes across today like a print version of old newsreel footage: a portentous voiceover signals urgency as jackbooted storm troopers occupy the Ruhr. Government agencies, the authors warned, were farming out various functions to high-priced consultants, secretive think tanks, and corporate vested interests — accountable to no one! And "outsourcing" was not the only issue. Some parts of the government, they went on, might even be sold off completely — turned into private businesses! The process was "cloaked in contractual and other formal approvals by the various executive departments," but make no mistake: it amounted to nothing less than a "drive to merge Government and business power to the advantage of the latter."

A little more than a decade later, the shadow government

was out of the shadow. There is a plausible rationale for privatization—one that often makes sense in the short run and for specific tasks. Private contractors may be able to operate more efficiently than government agencies do. Marketplace signals may prove to be more direct and powerful than bureaucratic ones. And why shouldn't the government hire outside specialists for help with certain chores, the way any household or business does? In the 1980s Ronald Reagan created a presidential commission on privatization to study not how the boundary between public and private might be bolstered but how it could be pushed out of the way even further, to give private interests more opportunity to move in. The same idea surfaces in the "reinventing government" movement taken up by the Clinton administration: "We would do well," one expert advised, "to glory in the blurring of public and private and not keep trying to draw a disappearing line in the water." Since then privatization has affected every aspect of American public life. It has also moved beyond narrowly defined tasks to embrace broad government functions. (It even seems to be crossing from the temporal realm into the spiritual one: because of a shortage of Catholic priests in America, requests for prayers made by Americans in their local churches are now being outsourced to Catholic priests in India, with a portion of the original donation remitted overseas.)

As noted, the most visible surge in government outsourcing has come in the realm of the military. In *The Soldier and the State*, written only half a century ago, Samuel Huntington observed, "While all professions are to some extent regulated by the state, the military profession is monopolized by the state." He could not write that sentence today. But it's not just the military; every facet of personal "security"—one of the basic elements of the social contract—is increasingly in the hands of private business. It was not until the nineteenth century that America's urban governments, by setting up local police forces, managed to make an ordinary person's daily safety a matter of real public responsibility. This was a major advance, though perhaps temporary. No one with money relies on such guarantees any longer (nor did they in Rome, where police forces as we know them were virtually nonexistent). More and more people have withdrawn into protected enclaves, a Latin American innovation successfully transplanted north. Private security is a major growth industry; in 1960 there were far more police officers than hired security guards in America, whereas today private guards outnumber the police by a margin of 50 percent. Individuals may owe nominal allegiance to a town or a state, but their true oath of fealty is to Securitas or Guardsmark.

One of the chief obligations of any government is simply to dispense justice — to resolve disputes, oversee legal business, mete out punishment. These functions were once held in private hands. After a stint as a public responsibility they are now migrating back. Lawyers and clients increasingly shun the civil courts — congested, expensive, fickle — and instead buy themselves some private arbitration, provided by a growing cadre of profitable "rent-a-judge" companies. As for the criminal-justice system, those sentenced to prison may very well do their time in a private facility, run on behalf of state and federal governments and operated by a company with some former public official in its management to grease the wheels. Faced with rising numbers of inmates (largely a consequence of "three strikes and you're out" laws), and unwilling to raise taxes to build more public prisons, governments at all levels have found that the easy, cost-effective way is to turn the prison industry over to the private sector: to a behemoth like the Nashville-based Corrections Corporation of America, or to one of many smaller companies. All but four states have contracted out some elements of their prison systems to private companies. Entrepreneurs known as "bed-brokers" help move prisoners from crowded facilities in one state to less-crowded ones in other states. In Rome, private companies sprang up to provide for the torture and crucifixion of troublesome slaves, relieving estate owners of the need to deal with the matter themselves. Crucifixion is one penal market niche that American companies have not sought to exploit.

America's public colleges and universities are fast losing their public character. These institutions were created under the terms of an act signed by Abraham Lincoln in 1862, which provided federal land grants to the states as a basis for public financing of higher education. But state support is diminishing. Nationwide, state legislatures are picking up only about two thirds of the annual cost of public higher education. For the University of Illinois, the figure is 25 percent. For the University of Michigan, it's 18 percent. The president of Penn State has warned of a "slow slide toward privatization." What makes up the difference in funding? To some degree, higher tuition. But even more it's money from private donors and private corporations, creating an incipient "academic-industrial complex" at public and private institutions alike. You can't escape the signs, whether small and symbolic or large and concrete (and brick and steel). At the University of California at Berkeley, one administrator is officially known as the BankAmerica Dean of the Haas School of Business. Most of the funding for Harvard's Center for Risk Analysis, which studies public-health issues related to food and the environment, comes from private industry, including chemical and pesticide manufacturers. Rice University has a Ken Lay Center for the Study of Markets in Transition, endowed by the late former chairman of Enron. Much money for universities comes with strings attached—most important, the power to push research in certain directions and perhaps away from others, and the ownership of patents deriving from sponsored research. The confusion between what's public and what's private starts with institutional self-interest and runs down through the interests of all the individuals involved. A professor at Wayne State University, who holds a chair in marketing that bears the Kmart name, is unabashed in his embrace of the patron-client relationship:

"Kmart's attitude has always been: What did we get from you this year? Some professors would say they don't like that position, but for me, it's kept me involved with a major retailer, and it's been a good thing."

Sociologists have a term for what is occurring: they call it the "externalization of state functions." Water and sewage systems are being privatized, and public hospitals and public-health programs. Voucher programs and charter schools are a way of shifting education toward the private sector. The protection of nuclear waste is in private hands. Meat inspection is done largely by the meatpacking companies themselves. Americans were up in arms when they learned that Dubai Ports World, a company in the United Arab Emirates, would soon be in control of the terminals at half a dozen major U.S. seaports — only to discover that terminal operations at virtually all American ports had long ago been privatized, and that 80 percent of them were already operated by foreign companies, the largest of which is Chinese. Serious proposals to privatize portions of Social Security have been on the table, and they will doubtless appear there again. One effect that privatizing Social Security would surely have is to encourage the affluent to opt out into private accounts to the greatest degree possible, eroding the communal sense of solidarity (as FDR understood) that has been the program's mainstay. Meanwhile, the new Medicare prescription-drug plan effectively puts an enormous government program into the hands of private insurance and drug companies. Those inclined toward dystopian fantasy should consider Bruce Sterling's futuristic novel *Holy Fire*. It conjures a society in which people are so beholden to private employers for the medical benefits that bring increasingly potent health care that they will do anything to avoid losing a job and therefore losing coverage — leading to a gradual reemergence of slavery in fact if not in name.

Many services that used to be provided free of charge now must be paid for — government by user fee. Detailed statistical data from the Census Bureau and other agencies was once available to everyone; now it's being sold, mainly for marketing purposes, and often at prices that only private corporations can afford. The vaults of the Smithsonian were once open to documentary filmmakers regardless of provenance and financing. Now an agreement between the Smithsonian and the cable company Showtime has created something called the Smithsonian Networks, which will have jurisdiction over, and priority access to, certain kinds of material. The documentary maker Ken Burns says of the deal, "It feels like the Smithsonian has essentially optioned America's attic to one company." State governments across the country are selling off such public assets as airports, highways, and even state lotteries to private companies, or handing them over under long-term leases.

Is there any government function that can't be transferred to some private party? A considerable amount of tax collection is now done, in effect, by casinos; rather than raise taxes to pay for services, legislatures legalize gambling and then take a rake-off, maybe earmarked for education or highways, from the profits earned by private casino companies. It's "tax farming" for the modern age, recalling the hated Roman practice of selling the right to collect taxes to private individuals (including the apostle Matthew in the Gospels) who were then allowed to keep anything over what they had agreed to remit to the government. (Allowing citizens to pay the IRS with credit cards amounts to the backdoor creation of another form of tax farming: credit-card companies turn into the government's tax collectors and reap a percentage of the tax paid until that credit-card debt is cleared.) As the recent revelations about torture have made clear, even official interrogations for national-security purposes have been outsourced—in this instance to other countries through the process known as "extraordinary rendition." The sale of naming rights for public facilities and other amenities attracts notice mostly for the ungainly nomenclature that results—mutants like the Mitsubishi Wild Wetland Trail, at the New York Botanical Garden, in the Bronx, and the Whataburger Field, in Corpus Christi. But it's also a sign of creeping privatization, as corporations, the Agrippas of our era, provide the kind of investment that the public purse can't or won't. Thus, to attract more corporate underwriting, the Department of the Interior has proposed that America's national parks be liberally opened up to the sale of naming rights. No one is suggesting that there will soon be a Weyerhaeuser Redwood National Park or a J. Crew Cape Cod National Seashore. But might there be a Fujifilm El Capitan Scenic Vista? A Sherwin-Williams Painted Desert Trailhead?

Contract Killings

THE PRIVATIZATION of power isn't a phenomenon of the margins, a footnote to history; it's a central dynamic of American public life. One study from the late 1990s suggests that the "privatization rate" — the rate at which public functions are being outsourced — is roughly doubling every year. On paper the federal work force nationwide, leaving the military aside, appears to total about two million people. But if you add in all the people in the private sector doing essentially government jobs with federal grants and contracts, then the figure rises by 12 or 13 million. It seems somehow not surprising that for the first time America has a president with an M.B.A. The commercialization of government probably explains why so many Washington entities are now referred to as "shops": "lobby shop," "counterterrorism shop." A private company that provides people to do government work of many different kinds is known as a "body shop." It may be that every instance of privatization can be defended on the merits — as efficient, thrifty, and wise. Surely many of them are. There's no question that in certain ways the private sector can outperform the public sector. Users of the *cursus privatus* — Federal Express, UPS, DHL — would sooner renounce citizenship than go back to relying only on the United States Postal

Service. The problem is the cumulative effect of privatization across the board—projected out over decades, over a century, over two—and the leaching of management capacity from government. This is the same *"mis*direction" of government force that MacMullen discerns in Rome: easier to observe in retrospect, when the whole film is available, than in the brief, real-time clip any of us is allowed to see.

An analyst at Johns Hopkins observes, "Contractors have become so big and entrenched that it's a fiction that the government maintains any control." One obvious recent example is the rebuilding effort in Iraq. To supply the army or provide other services, traders and contractors often traveled with Roman legions; Julius Caesar had such a person with him during the Gallic Wars, explicitly "for the sake of business." There may have been no alternative to giving the reconstruction job in Iraq to private corporations, including giant combines like Halliburton, but the result has been an effort that defies management or accountability. Private companies are exempt from many regulations that would apply to government agencies. They can use foreign subsidiaries to avoid laws meant to restrain American companies. (Before the war, Halliburton itself used subsidiaries to do business with Iran, Iraq, and Libya, despite official American trade sanctions against all three countries.) The records of private companies can't be obtained through the Freedom of Information Act. Congressional committees aren't set up to look over the shoulders of private companies, the way they do with government departments. The evidence of widespread corruption in the Iraq rebuilding effort is beyond dispute. The rigid impartiality of many contractors has been more easily disarmed than the insurgents. Companies have overcharged the government by tens of millions just for food and gasoline. The Coalition Provisional Authority official in charge of disbursing some $80 million in one Iraqi city turns out to have been at the center of a bribes-and-contracts conspiracy: "a maelstrom of greed, sex, and gun-running," according to one account—the kind of tale that might have come from Lepcis. But the larger issue is that orders aren't followed "along the whole train of power"; there isn't any train of power—just a collection of private interests. A source quoted by the Special Inspector General for Iraq Reconstruction, which issued a scathing report at the end of 2005, said simply, "The CPA was contracting, companies were contracting subcontractors, and some people who didn't have authority such as the ministries were also awarding contracts."

This is the pattern everywhere in government, even when the most sensitive matters of national security are involved. More and more secret intelligence work—translation, computing, interrogation, analysis, reporting, briefing—is being farmed out to private entities. Not only is the intelligence community becoming further fragmented, but, because the new jobs pay so well, a "spy drain" is drawing officers out of the public sector and into the private market. And the drain isn't restricted to spies: at least ninety former top officials at the Department of Homeland Security and the White House Office of Homeland Security are now working for private companies in the domestic-security business. In a recent novel the former CIA officer Robert Baer describes what has become business as usual:

> Everyone I knew seemed to be doing it once they hit the magic fifty: Retire on a Friday, back in the building Monday morning with a shiny new green badge ... They doubled their salaries overnight, while the companies that hired them got experts trained on the taxpayers' tab and a straight shot into the vitals of the CIA, where they could work on landing more contracts.

The company that fits this profile more than any other is known by the acronym SAIC, which stands for Science Applications International Corporation. Founded in 1969 to take on a small number of specific government research projects, the company has grown into something that might as well be a full-fledged government agency, with a staff of 43,000—except that it operates in its own interest and exists outside ordinary government oversight. Its annual intake from the federal government exceeds $6 billion. In some ways it is the quintessential "body shop" — it will take on any project the government wants to outsource — though "brain shop" might be a better term, because most of its work involves expertise that the government no longer has, or perhaps never did have. Its senior executives tend to be drawn from the government agencies it does business with, and they often cycle back into government for a period of time — in other words, the people who define the need are also satisfying the need. SAIC is managing preparations for the Yucca Mountain nuclear-waste repository. It has designed data-mining programs for the NSA and managed computer systems for the Immigration and Naturalization Service. It was brought in when the CIA wanted to monitor private use of the Internet by its employees. After the invasion of Iraq, SAIC established a newspaper and a television station in Baghdad. Many of the efforts that SAIC has been involved in have been white elephants. The $100 million computer system it set up for the FBI had to be abandoned; so did its $1.2 billion Trailblazer system for the NSA. The media venture in Iraq, ludicrous on its face, went nowhere. But SAIC continues to seek and win federal contracts: it currently holds 10,000 of them, all told. Its appetites are indiscriminate, pursuing whatever is available simply because it is, and ultimately for SAIC's own ends. Ramsay MacMullen's comment about effective government functioning comes to mind — how the "train of power" must link authority at the center with public purpose at the margin: "At every point of connection the original intent must be transmitted as it was received. Otherwise it will come to nothing." One former executive has said about SAIC, "We used to joke that it really was Kentucky Fried Chicken consulting."

Meanwhile, the government is turning the job of border police over to multinational contractors, a task that will in turn be subcontracted out to dozens of smaller companies, further attenuating any attempts at government control. Lockheed Martin, Raytheon, Boeing, and Northrop Grumman are among the corporations that indicated they would submit bids to build a high-tech "virtual fence" along the Mexican border, with an array of motion detectors, satellite monitors, and aerial drones. A Homeland Security Department official conceded the abdication of government leadership, saying to the companies, "We're asking you to come back and tell us how to do our business."

The activities of government are, in effect, being franchised out. You can't help lingering over the concept of "franchise," wondering what a latter-day Geoffrey de Ste. Croix would make of it. Like *suffragium,* the word originally had to do with notions of political freedom and civic responsibility; it comes from the Old French word *franc,* meaning "free." Because free people were those who had been granted certain rights, the word came to be associated with the most fundamental political freedom of all: to exercise your franchise is to exercise your right to vote. Only much later, in the mid twentieth century, did the idea of being granted "certain rights" acquire its commercial connotation: the right to market a company's services or products, such as fried chicken or Tupperware. Today, to have a franchise on something is in effect to have control over it. Kellogg Brown & Root has a franchise on the building of military bases. SAIC seems to have a franchise on setting up government computer systems. The English dictionary published under the auspices of Wikipedia may not be the most scholarly or authoritative reference book, but because of the way it's compiled, with entries from people at large, it's the one that comes closest to reflecting popular understandings of language at any given moment. In the Wiktionary, the commercial meaning of "franchise" is now the primary definition. The definition involving political freedom and the right to vote comes fifth.

Looking back at the history of "franchise," then, it's tempting to write this epitaph: Here, in miniature, is the political history of America.

4

THE OUTSIDERS
When People Like Us Meet
People Like Them

When given the German command he went out with the
quaint preconception that here was a subhuman people
which would somehow prove responsive to Roman law. . . .
He therefore breezed in — right into the heart of Germany
— as if on a picnic.

— Velleius Paterculus,
on the disaster at Teutoburg Forest

What struck me most about the palace was the completely
self-referential character of it. It was all about us, not about
them. People would walk around the palace with a mixture
of venal and idealistic motives. None of them knew Iraq.

— an American diplomat inside Baghdad's Green Zone

I N THE YEAR 15 A.D. the legions commanded by the Roman general Germanicus — stepgrandson of the recently
departed emperor Augustus, adopted son of the new emperor
Tiberius, father of the future emperor Caligula — picked their
way deep into the thick forests and forbidding marshes east of
the Rhine, looking for the spot where, six years earlier, a terrible
slaughter had taken place. The scene is set by Tacitus in the first
book of his *Annals*. Germanicus had been waging a campaign
against one of the German tribes when, on an impulse, he set
out with his army to discover the battlefield known to history as

Teutoburg Forest. There, in 9 A.D., three legions commanded by Publius Quinctilius Varus had been tricked and then ambushed by a smaller German force and virtually annihilated. The Germans had also made off with the three silver legionary eagles, which symbolized the honor of each legion and the power of Rome — a profoundly shameful event. It was the most humiliating defeat ever suffered by a Roman army. The repercussions were staggering, and not only for Rome. We live with them to this day.

Here's Tacitus:

> The dismal tract [was] hideous to sight and memory. Varus' first camp, with its broad sweep and measured spaces for officers and eagles, advertised the labours of three legions: then a half-ruined wall and shallow ditch showed that there the now broken remnant had taken cover. In the plain between were bleaching bones, scattered or in little heaps, as the men had fallen, fleeing or standing fast. Hard by lay splintered spears and limbs of horses, while human skulls were nailed prominently on the tree-trunks. In the neighboring groves stood the savage altars at which they had slaughtered the tribunes and the chief centurions.

Germanicus ordered his men to gather the remains of the fallen into a funeral pyre. When the fires had been lit, he said, "Let us be soon gone from here, and let the mists of time cloud its very existence."

And so they left. Not long afterward Germanicus would fight a series of inconclusive battles with the very German forces that had humbled Varus. A few years later the leader of those Germans, a chieftain named Arminius, was killed during some sort of family squabble. But the mists of time have not clouded the memory of Teutoburg Forest, because the central lesson is perpetually compelling.

On its mental map Rome pictured itself as all-important, all-

knowing, all-powerful. The inevitable corollary to this perspective was a view of outsiders as often unfathomable, certainly
inferior, and in any event not worth the bother of trying to understand. The great physician Galen once wrote, "I am no more
writing for Germans than for wolves and bears." Such an outlook is typical of empires. And for Rome, much of the time, an
oblivious frame of mind did not really matter: Roman power was
overwhelming, and the fear it instilled could prove as effective as
actual force. Sometimes obliviousness doesn't matter for America, either. But often — and increasingly — it matters a great deal.
How people in the outside world think and behave, and respond
to America's thinking and behavior, are variables that must be
taken into account. They are as important as objective factors
like the strength of an economy and the size of an army. Smug-
ness or indifference can prove catastrophic. It did for Varus.

Four Days in September

AFTER THE PILGRIMAGE of Germanicus, the Teutoburg battlefield was lost to history for nearly 2,000 years. The funeral pyres
burned down to ash. Trees and marshes reclaimed the killing
ground. The true location of Teutoburg Forest was the subject of learned — and often misguided — speculation for centuries (a monument stands at the wrong place), but the actual site
was not in fact rediscovered until the mid-1980s, when a British officer stationed in Germany, Tony Clunn, by avocation an
archaeologist, turned up some Roman coins in a remote field
near the town of Kalkriese, near Osnabrück, in Lower Saxony.
He had been looking for the Teutoburg battlefield for years, was
deeply knowledgeable about the literature and the terrain, and
had decided to play some hunches. One of them paid off. The
coins at Kalkriese led to more coins, and then to military equipment and personal items like keys and hairpins and cloak clasps.
Eventually the earth divulged the horrifying scale of the calam-

ity that had befallen Varus and his army: the debris field from the protracted battle covered an expanse of ground three miles wide and four miles long.

P. Quinctilius Varus was a man typical of his class and his time. He was a lawyer, connected by marriage to the imperial family, and had served as governor of the provinces of Africa and Syria, where he seems to have ably used his office to enrich himself. He was not of a temperament to deal patiently with those he regarded as troublemakers or inferiors. Once, in Syria, he ordered the crucifixion of 2,000 insurgents to quell an outbreak of unrest. (It worked.) Assessments of his competence are mixed. A man who knew and disliked him, Velleius Paterculus, described Varus as "somewhat ponderous in mind and body." But when he was appointed governor of Germania, there was little reason to believe that disaster lay ahead. By steady accretion the power of Rome had advanced to the Rhine and the Danube, and the next logical step was to incorporate the lands that lay beyond: "free Germany," what is today the German heartland, all the way east to the Elbe. Roman advance encampments had already been planted. Roads had been cut. Relations with some of the German tribes appeared to be stable. No one had suggested to Varus that he would, in effect, be greeted with flowers, but he shared the conviction that the Roman occupation of this prospective new province was a done deal; that the Germans were manageable; and that the task ahead was mainly to garner revenue, dispense justice, and otherwise administer this territory, which would soon be absorbed formally into the imperial system. The moment for nation building was at hand. "The Germans," Paterculus writes, "a race combining maximum ferocity with supreme guile (and being born liars besides), fawned upon Varus, making much of their lawsuits, marvelling at his jurisprudence and flattering him regarding his civilizing mission."

As the summer of 9 A.D. neared its end, Varus led his force — some 15,000 legionaries and an equal number of camp followers

—from the north of Germany toward winter bases on the Rhine, secure in a presumption of superiority to any enemy. But there was one matter to attend to. Varus had been told by Arminius, a prince of the Cherusci, about a small uprising by one of the German tribes, and he resolved to make a detour and put it down. Another German chieftain warned Varus not to trust Arminius. Varus paid no heed. Had not Arminius served bravely with Rome? Even been made a citizen and given honors?

At the time, much of Germany was densely covered with woodland, and clearings were as likely to be sodden bog as dry land. The opening scene of *Gladiator*, with its muddy fortifications and gnarly trees, vividly captures this unfriendly environment—so different from the open countryside and broad vistas that the Romans knew from the Mediterranean. Arminius led Varus deep into unknown territory. "The Romans were having a hard time of it, felling trees, building roads, and bridging places that required it," the historian Cassius Dio writes. "They had with them many wagons and many beasts of burden as if in a time of peace; moreover, not a few women and children and a large retinue of servants were following them." Military planners today speak of a "teeth-to-tail" ratio—the ratio of fighting personnel to ancillary support personnel. Americans have a very long tail, and the Romans did too—a significant encumbrance in a guerrilla war or an insurgency. The Roman lines thinned out along the narrow trackways, stretching for miles and miles. Fate proved unkind: violent rains set in, turning fields into marsh, and marsh into swamp. "While the Romans were in such difficulties," Cassius Dio goes on, "the barbarians suddenly surrounded them on all sides." The Romans were about to get a lesson in what today would be called asymmetric warfare. For three days Varus and his legions held out, defending themselves against the barbarian attacks while trying to retreat toward the nearest Roman base, at Haltern.

There is a computer game called Rome: Total War, in which

a player can theoretically play the role of Varus and win. The real Varus could not. On the fourth day, the Romans were finally overrun. Those who were not killed immediately were killed eventually, after torture. Some 30,000 people died in all. Before the end, P. Quinctilius Varus fell on his sword, as would have been expected of any noble Roman in the face of such dishonor. Only a handful of survivors made it safely to Haltern.

It took weeks for news of the catastrophe to reach Rome. The impact is hard to overstate. Imagine a combination of 9/11, Pearl Harbor, and Little Bighorn. In a single day three entire legions — the XVII, the XVIII, and the XIX, representing some ten percent of Rome's invincible military — had been wiped out. The Romans could never bring themselves to use those legionary numbers again. It was all the worse for coming at a moment saturated with feelings of omnipotence: Rome had been about to hold a formal triumph to celebrate a series of military victories in the northern Balkans by Tiberius, the adopted son and presumptive heir of Augustus. Instead came word of the slaughter in Germany. Before long Augustus would also receive the head of Varus, severed by the Germans and forwarded through an intermediary for delivery into his hands.

Were the Germans on the march? Would Gaul be next? Would the defeat embolden other enemies? Was the capital itself at risk? Public paranoia was inflamed by omens — the red and orange "threat level" warnings of their time. The Temple of Mars had been struck by lightning! Locusts had flown into Rome and been devoured by swallows! For decades German auxiliaries had served as loyal soldiers in Rome's armies. Now, suddenly, they were viewed with suspicion. Augustus disbanded the German cavalry that had long helped to protect him. He sent troops into neighborhoods of Rome where German immigrants lived. He gave emergency powers to governors in far-flung provinces. He compelled free citizens to join a new armed force, which he sent north to the Rhine.

In the aftermath of Teutoburg Forest the Romans forever lost their taste for ambitious expansion into free Germany. Yes, they would trade with the various German tribes, and put them to work, and attempt to manipulate them diplomatically, and absorb some of them into the armed forces, and if necessary fight them, but henceforward the northern Roman frontier would remain where it had been: more or less along the Rhine and the Danube. That frontier traces a cultural boundary to this day, a dividing line between languages, liturgies, and eating and drinking habits. Europeans call it the "oil-butter line." You could also think of it as the "wine-beer frontier." Arminius himself is seen with some justice as a precursor of an independent Germany, and his memory lives on in the common name Hermann. The name and the memory came to America with immigrants. An enormous monument to Arminius, erected a century ago by a group of German Americans called the Sons of Hermann, stands in a park in New Ulm, Minnesota. It is a city of brew pubs, not wine bars.

What conclusions did the Romans draw from their defeat? It's a hoary truism that people don't learn the lessons they most urgently ought to heed. Some Romans blamed sheer incompetence by Varus, some the bad weather, some a supernatural judgment. No one at the time hinted at the true explanation, the strategic premise of the disaster, which was this: the Roman disinclination either to understand the minds or to credit the capabilities of people unlike themselves. "Underestimation of space was matched by the under-rating of people," one historian concludes. Another writes, "The Romans simply could not believe that their military forces had been outfought by the northern barbarians." Even if evidence of barbarian technological skill and organizational precocity had been presented in advance, he goes on, "Augustus and other Roman officials would not have been receptive."

According to Suetonius, after receiving news of the battle

Augustus could be heard hitting his head against a door and lamenting aloud, "Quinctilius Varus, give me back my legions!"

Reality Check

SPARTACUS MAY NOT be a historically fastidious movie, but it captures an actual state of mind. Right after the campy bath scene, in which Crassus (Laurence Olivier) is washed by the slave boy (Tony Curtis), Crassus takes the slave out to the balcony and shows him the legions passing by. "There, boy, is Rome — there is the might, the majesty, the terror of Rome. There is the power that bestrides the world like a colossus. No man can withstand Rome, no nation can withstand her — how much less a boy?" When I came upon those words recently, it was hard not to recall a remark made by a Bush administration official to the reporter Ron Suskind: "We're an empire now, and when we act, we create our own reality. And while you're studying that reality — judiciously, as you will — we'll act again, creating other new realities, which you can study too, and that's how things will sort out. We're history's actors . . . and you, all of you, will be left to just study what we do."

Among the people they ruled the Romans aggressively displayed the symbols of their power — like the well-known fasces, the wooden rods bundled around an ax and tied up with red straps. The fasces, once carried before consuls, eventually were carried before emperors; they were dipped ceremonially to show respect, as you dip a flag. They might be carried in reverse at a funeral, as boots are placed backward in the stirrups in a modern military cortege. To seize hold of the fasces was to proclaim an attempt at seizing power. As symbols, the fasces today seem relatively innocuous, the bundled rods often given the anodyne interpretation "strength in unity." They were adopted as a republican symbol by a young America. Look behind the president when he gives the State of the Union address, and you'll see fasces on the wall of the House chamber. You'll see them on the massive marble seat Abraham Lincoln occupies in his memorial. But historians remind us what the fasces originally were: "a portable kit for flogging and decapitation." At one time they were widely used as such. Carried before provincial governors, they exuded an "aura of latent violence" and were deeply resented by many non-Romans. Imagine if, instead of our flag flying over the entrance to an American embassy, you saw a bas-relief of stun guns and M-16s.

The fasces epitomized a heedless arrogance that sometimes led to needless trouble. Punctuating a long history of great Roman military triumph stand dark moments of great military disaster. Behind virtually all of them lay an attitude that was "almost never cautious and often verged on the reckless," as one historian concludes. Another, offering a catalogue of examples, cites "the old aggressive culture of the Roman army — eager to fight, impatient with tactics."

One of those disastrous moments involved that same Crassus played by Olivier — Marcus Licinius Crassus, the man who in 71 B.C. put down the slave revolt that the gladiator Spartacus had led, and crucified some 6,000 of the insurgents along the length of the Appian Way. Crassus was rich, and known to care deeply for his money. He controlled an extensive network of patronage, and became a member of the first triumvirate, with Caesar and Pompey. Unlike the other triumvirs, he had never been given credit for a true military victory against a worthy enemy — the slaves were seen as mere vermin — and he had never been awarded an official triumph. And so, in 53 B.C., "desiring for his part to accomplish something that involved glory and at the same time profit," as Cassius Dio explains, he marched hastily and overconfidently against the Parthians, Rome's longtime eastern foe. Crassus crossed the Euphrates with intelligence compromised both by its source and by his own wishful thinking — by an inclination, as a later age might say, to misunderestimate the capacities of the enemy. The Parthians were not the equal of the Romans in any strict military sense, but as horsemen in mobile warfare they were superior. Their ability to skillfully shoot arrows behind them while retreating at a gallop gave rise to the expression "Parthian shot," meaning the last word in an argument, a decisive putdown. Crassus and his conventional forces were teased and bedeviled by the canny Parthians, and drawn deeper and deeper into an alien environment. Finally, at Carrhae, in what is now southeastern Turkey, his army was destroyed.

The Parthians were well aware of the Roman general's love of money. "And not only the others fell, but Crassus also," writes Cassius Dio, "either by one of his own men to prevent his capture alive, or by the enemy because he was badly wounded. This was his end. And the Parthians, as some say, poured molten gold into his mouth in mockery." What the Parthians certainly did do was to conduct a mock parade of the hated fasces, with the heads of Roman soldiers impaled upon them.

A century and a half before Carrhae, in 216 B.C., during the Second Punic War, there was the disaster at Cannae, when, again, Romans with superior numbers proved "impetuous" (the word used by the historian Polybius), abandoning all caution and falling upon Hannibal and his Carthaginians, only to lose eight legions to the stratagem of "double entrapment," now taught at war colleges everywhere as the oldest trick in the book. The Roman forces were led by two consuls. One of them, the seasoned Aemilius Paulus, seeing that the flat terrain gave advantage to Hannibal's cavalry, urged prudence and delay. The other, the green and hotheaded Terentius Varro, who enjoyed the support of the rank and file and of public opinion in Rome, wanted to attack at once. Because the consuls alternated command from day to day, Varro needed only to let the earth spin to get his way. He led the Romans into the Carthaginian center and "advanced so far that the Libyan heavy-armed troops on either wing got on their flanks" and closed in for the kill. It was a defeat the Romans would never forget. And yet strategists in the capital entertained no second thoughts about what one modern analyst calls Rome's culture of aggression: "Nor did the Romans entirely learn their lesson at Cannae," he writes. "Subsequent Roman armies still threw themselves upon Hannibal and kept being defeated."

The Romans recovered from Cannae, but they never really recovered from the Battle of Adrianople, in 378 A.D., when the emperor Valens sought to contain a large force of Visigoths that had been allowed to cross the Danube in an area west of Constantinople and was now wandering about, hungry and dangerous. Valens had a substantial army with him, but was counseled by one of his generals, Sebastian, to avoid a frontal attack: with patience, the Goths could be harried and starved into surrender or retreat, as had often happened before. At the very least, Sebastian urged, why not wait for reinforcements from an additional Roman army that was on the way? The emperor would have none of it. According to Ammianus Marcellinus, Valens was seized with "a kind of rash ardor" and "determined to attack them at once." He was emboldened in part by intelligence reports that had underestimated the number of Goths in front of him. Valens was also miffed: he had been insulted by the low rank of the barbarians who came to treat for peace. It may be, too, that he did not wish to share the glory with another commander and another army. As the chronicler Josephus wrote in a different context (explaining why, centuries earlier, the Roman general Titus had chosen to storm Jerusalem rather than let it come into his hands by a prolonged siege): "Time would accomplish anything, but for glory speed was necessary." Fundamentally, though, the decision came down to the fact that Valens did not take the Goths seriously as military opponents. In the words of a modern analyst, "All the Roman commanders, with the possible exception of Sebastian, acted with the typical arrogance of a well-equipped, 'civilized' army dealing with what they saw as rabble."

The battle was joined, and the Romans, hemmed in by Gothic cavalry and blinded by brush fires, lost the freedom to maneuver. The rout was total. The body of the emperor was never found. Ammianus Marcellinus concludes, "The annals record no such massacre [in battle] except the one at Cannae."

A Clash of Caricatures

THE ATTITUDE OF ROMANS toward non-Romans is not simple, and it's not one-dimensional. For Americans, though, it does have familiar elements, starting with a sense of "exceptionalism," of having been chosen for a special purpose. That word was first applied to Americans by Tocqueville, but the outlook it describes goes back to the moment of settlement. "Wee shall be as a Citty upon a Hill," wrote John Winthrop, the governor of the Massachusetts Bay Colony, in 1630. "The eyes of all people are upon us."

If they were arrogant, the Romans were also, like Americans, unabashedly syncretic — they borrowed heavily from others. They had come into their own while inhabiting a world in which certain foreign peoples — Greeks, Phoenicians, Egyptians — were by most cultural measures far more advanced than they were. As Rome expanded, it took these groups under its sway and adopted what they had to offer. "The Romans," one historian writes, "thought of themselves not as a single ethnic group but as embracing all people." It would take centuries, but eventually free inhabitants throughout the empire were granted Roman citizenship. The dress and grooming of faraway peoples, including even the barbarians, often enjoyed a Roman vogue — you see it in statuary and in the images on coins. The emperor Caracalla was named for the *caracallus,* the Gallic cloak he habitually wore. As ethnic pragmatists, the Romans would probably have understood certain aspects of modern America — for instance, the introduction a few years ago of a new picture of "Betty Crocker" on boxes of mashed-potato mix, to better reflect the American demographic reality. (A computer used images of several dozen diverse women to create a darker and more ethnically resonant Betty, in contrast to the original pallid, blue-eyed version of 1936.)

At the same time, the Romans really did see themselves as distinct from, and superior to, non-Roman peoples, both those within and especially those beyond the pale of empire. Think schematically in terms of concentric circles. In the center is the city of Rome and its citizens. Next come others on the Italian peninsula, who only gradually are permitted to share in Roman citizenship and governance. Then come the freeborn among the conquered peoples. Finally, at the edge, in the thin halo remaining between the imperial frontier and the outside limit of the known world, lies *barbaricum,* the mysterious and forbidding barbarian domains. The word "barbarians" itself connotes incomprehensibility — it's generally said to be an onomatopoeic term that originated with the Greeks, to whose ears the speech of outlanders sounded like a meaningless "bar-bar-bar." One way of translating it would be as "gibberish people." It is certainly true that in a broad zone along the frontiers, Romans and non-Romans knew one another intimately; Roman perceptions, whether friendly or hostile, would not have been the stuff of distant caricature. But personal attitudes of one kind ("some of my best friends are Quadi") easily coexist with generic stereotypes of another. Fear only makes things worse. You find a lot of sentiments like the following, from an anonymous writer in the fourth century A.D.: "Wild nations are pressing upon the Roman Empire and howling round about it everywhere." Or this, from Jerome in the early fifth century: "May Jesus protect the world in future from such beasts!"

The Romans needed to be aware of the activities of specific groups, but unlike the Greeks — inveterate explorers and writers of travelogues — they didn't have a burning curiosity about the unknown, and they didn't methodically try to penetrate and understand neighboring cultures for national-security or any other purposes. Foreign "intelligence" was often haphazard — the result of serendipity rather than design. Ethnographic works like the portrait of the Germans and their culture by Tacitus, with all its insights and biases, are the great exception. The historian Susan Mattern notes that the Romans derived foreign policy from their own values and cultural identity, rather than from cold cost-benefit calculations. In the same vein she adds, "Their decisions were based more on a traditional and stereotyped view of foreign peoples than on systematic intelligence about their political, social, and cultural institutions." To be sure, the Roman military took pains to gather whatever tactical information it could along the frontiers. Much information arrived willy-nilly, with a delegation from here, with some hostages from there, or on the lips of a trader who'd heard something from another trader. One historian writes, "The Romans acquired information that we would deem vital for the conduct of foreign relations only randomly through a variety of ad hoc sources" — the ancient equivalents of, say, CNN, Ahmed Chalabi, and "friends in the oil business." In hindsight it's fascinating to watch as the Romans try to figure out the ultimate catalyst for a rash of barbarian invasions in the late fourth century. What's making so many different groups suddenly spill across the Rhine, the Danube? It takes the Romans quite a while to infer the existence of the Huns, still thousands of miles away, out on the far steppes, a place that was *terra incognita* — just like the mind of radical Islam to most Americans before 2001 (and maybe still).

Intelligence collection was hampered partly, of course, by the general problem of communications. As good as Rome's were for its day, they were slow. The overland trip from Rome to Antioch and back probably took two months; the fast boat from Italy to Alexandria and back, under ideal circumstances, took three weeks. Ironically, in an age of instantaneous communications, American intelligence often suffers from a time lag just as significant. American analysts amass far more information than they can digest. The FBI office in Phoenix sent a warning to Washington in June of 2001, noting that an "inordinate number of individuals of investigative interest" were signing up for lessons at American flight schools. The warning might as well have gone by *cursus publicus* from Rome to Antioch: no one saw it for more than two months, until after the 9/11 attacks proved it germane. Again: CIA and FBI agents met in New York in June of 2001 to discuss an individual who turned out to be one of the 9/11 hijackers, and who had been raising eyebrows for months. But crucial information was not shared, and no one was looking for this individual when he entered the country shortly thereafter. Bureaucracy is the new geography.

The Romans established a whole taxonomy of non-Roman cultural traits and stereotypes, few of them flattering. The Greeks, of course, had been admired for their art, their literature, and their philosophy, but admiration can curdle into resentment, and Romans who affected Greek ways, in the manner of Americans who affect Anglicisms after a stint in London, were sometimes derided as "Greeklings." Other groups fared less well. The indigenous people of Iberia were said to brush their teeth with urine. The northern Gauls were seen as immoderate in their appetites, generally undisciplined, and without stamina in battle — an apparently durable stereotype. Syrians were notoriously dishonest, and in Cicero's view were "born slaves." The Egyptians were weak and degenerate, the Parthians dangerous and decadent. Jews were regarded as peculiar, standoffish, exclusive, and of course inferior; it must have been something of a surprise that these people, when they revolted in the first century A.D., would tie down the Romans for four full years and require more troops than the conquest of Britain. The free Germans, the ones living outside Roman reach, were implacably warlike and notoriously unsusceptible to notions of delayed gratification. Tacitus, who could have had a career at *The Economist,* offered this crisp, back-of-the hand dismissal:

> A German is not so easily prevailed upon to plough the land and wait patiently for harvest as to challenge a foe and earn wounds for his reward. He thinks it tame and spiritless to accumulate slowly by the sweat of his brow what can be got quickly by the loss of a little blood.

And so on. But let it not be said that the Romans thought themselves superior only by comparison with the defectives all around. Cicero was quick to point out good qualities in non-Romans, and to suggest that the Romans had risen to the top because they enjoyed the sanction of heaven itself: "Spaniards had the advantage over them in point of numbers, Gauls in physical strength, Carthaginians in sharpness, Greeks in culture, native Latins and Italians in shrewd common sense; yet Rome had conquered them all and acquired her vast empire because in piety, religion, and appreciation of the omnipotence of the gods she was without equal." Call it the idea of "Roman exceptionalism" — a shining city upon seven hills. You can't miss an echo in the religious righteousness of our own day — in the words, for instance, of Lieutenant General William G. Boykin, explaining the American capture of a Muslim warlord in Somalia in terms of Christianity's superiority to Islam: "I knew my God was bigger than his. I knew that my God was a real God and his was an idol."

You hear modern chords not only when the Romans speak about non-Roman peoples, but also when non-Romans speak about Rome. It was not all negative: Rome had plenty of Alistair Cookes. There would always be rebellions — they are as prominent a feature of *The Decline and Fall of the Roman Empire* as the elegant rumble of Gibbon's jokes — but Rome brought unprecedented peace and stability to the lands it ruled. The advantages of the imperium were undeniable, and in many ways Rome's yoke was light. Most religions were tolerated, as long as proper sacral obeisance was paid to the emperor (a problem for Jews and Christians). The local ruling class usually remained the local ruling class. Piracy was suppressed, and commerce flourished. Water flowed to areas that had once been dry. Strabo and Plutarch, in the first century A.D., were Greeks who traveled within the highest social circles. Both had visited Rome; Strabo actually lived there for many years. You'll find plenty of trenchant criticism of Rome and Romans in their writings, but the overall assessment is positive: Strabo and Plutarch bought into the Roman system. So did Appian, another Greek, from Alexandria. In his *Roman History* he extols the pluckiness of the Romans, marvels at their willingness to lavish wealth on godforsaken places ("on some of these subject nations they spend more than they receive from them"), and sums up two centuries of imperial rule as follows: "In the long reign of peace and security everything has moved toward a lasting prosperity."

Since the middle of the twentieth century, America has been seen by many outsiders as playing much the same role — perhaps to their annoyance, but also to their relief. America has done so, like Rome, for reasons of national self-interest; unlike Rome, it has done so without asserting actual sovereignty over the countless multitudes who receive some collateral benefit. America's advances in communications and medicine have spread everywhere. Its power has shaped the global infrastructure of security, finance, and trade. The world's 76,000 daily commercial flights navigate through the skies with the aid of American satellites. American fleets keep the sea-lanes safe. The world's computers operate mainly on American software, and the world's *lingua franca* is America's television and advertising English, not the English of Henry Higgins or Noel Coward. The breakfast buffet at the InterContinental Dubai flies many culinary flags: it offers miso soup and sushi, croissants and baguettes, rashers and kippers, dates and figs, porridge and cereal — an implicit showcase of that city-state's emergence as a mercantile crossroads, a latter-day Palmyra. But more than that, the breakfast buffet is an edible monument to the international order fostered by the country responsible for the Cheerios and Special K.

During the past five years, as America has confronted the hostility not only of radical Islam but also of many onetime friends and allies, the question "Why do they hate us?" has been asked and answered again and again. It's not because they don't know us: Foreigners seem to know a lot more about Americans than we do about ourselves. Every few years some new national survey laments the basic ignorance of American teenagers about the nation's past; high school students in Kiev and Kathmandu, it always turns out, are better informed about the Civil War and the New Deal than students in Trenton or Omaha. According to recent studies, Europeans are considerably more likely than Americans to have heard about Guantánamo and Abu Ghraib. Whether informed or not, the menu of complaint is well known: Americans are insensitive vulgarians—loud and uncultured, God-drunk and materialistic, implacable and self-congratulatory, blind to the needs and views of others—who wield their power with unthinking brutality and childlike clumsiness, largely in the service of oppressive regimes, hegemonic ambitions, and insatiable appetites. Or something along those lines. The novelist Margaret Drabble has admitted that anti-Americanism "rises up in my throat like acid reflux." Many decades ago, in an age of more artful savagery, the British diarist and diplomat Harold Nicolson told a friend that Europeans were "frightened that the destinies of the world should be in the hands of a giant with the limbs of an undergraduate, the emotions of a spinster, and the brain of a peahen." He forgot to mention one other trait: an ear of tin, as when Zbigniew Brzezinski, the former national security advisor, refers publicly to America's friends and allies as our "vassals and tributaries."

Valid criticisms lurk here, and so, of course, does a caricature, which Americans themselves can't help recognizing. I remember experiencing, as an American teenager living in Dublin and more or less totally assimilated to Irish folkways, a dread of summer, when wagon trains of tour buses would deposit Americans everywhere, with their loud shirts and loud personalities — a dread mixed even then with shame at my pathetic disloyalty. At any moment, I half expected, the cock would crow.

Anti-American sentiment is by now a permanent part of the ecosystem, turning up in toxic levels almost everywhere. Opinion surveys show America's positive image in the world to be eroding even among traditional friends — a development so familiar as to barely register as news. The words *"Civus Romanus sum"* — "I am a Roman citizen" — were once both a boast and a form of protection throughout the known world; during the 2006 World Cup, the American soccer team was the only one whose bus, for security reasons, bore no markings of nationality. Only occasionally does anti-Americanism take a truly startling turn, as when a columnist in our stalwart ally Britain — Athens to our Rome! — contemplated the impending re-election of George W. Bush and offered the comment "John Wilkes Booth, Lee Harvey Oswald, John Hinckley Jr. — where are you now that we need you?"

The Romans, too, presented a tempting target. To Christian writers, Rome was the Great Whore, just as to Islamists, America is the Great Satan. The Romans talked loudly of liberty — for themselves. The word hardly applied to those who came under Rome's sway. The Romans made a habit of publicly degrading the captured leaders of vanquished peoples, a practice now turning up unexpectedly in the American repertoire — recall the official video of Saddam Hussein's medical exam, broadcast worldwide. Roman envoys, serene in their arrogance, spoke contemptuously to foreigners, even to royalty, and in the expectation that all requests would be granted. The story is told of one Roman envoy who, wanting action from a foreign king, uttered no word of threat but with a stick simply drew a circle around the king's throne, in front of his people and courtiers, and then advised him not to pass outside the circle until he had an answer for the Roman Senate.

The barbarian tribes were illiterate, and left few direct impressions of their views of the Romans. The Greeks were another story. In Greek eyes, the Romans were cruel (not a personal trait for which ordinary Americans get much criticism) and rapacious, stripping lands of their culture and resources (a little more recognizable). The Greeks openly admired the Romans for some qualities — orderliness, rationalism — but they weren't endorsing the whole package. Polybius, a Greek who held many aspects of Rome in open esteem, nevertheless in one passage explicitly refers to the Romans as barbarians.

In the view of foreigners, the Romans could be laughably uncouth. Once, in the 1980s, when the annual Teamsters convention was held at Caesars Palace, in Las Vegas, the three-hundred-pound Jackie Presser, a candidate for union president, was carried into the convention hall on a gold litter by four men dressed as Roman centurions, to the piped-in chant of "Hail Caesar." The ancient Romans sometimes came across in much the same way. The story is told of one wealthy Roman who after the sack of Corinth ordered some antique works of art sent home, warning the shipper that if any were damaged he'd have to replace them with new ones. (Charles Foster Kane would have done the same thing.) The Greek philosopher Demonax, grown tired of listening to someone boast about how the emperor had made him a Roman citizen, delivered the Parthian shot: "A pity he did not make you a Greek."

The "Sameness" Delusion

ONE NOTABLE CONSTANT in American history is our lack of awareness of the rest of the world — or, if we're aware, our indifference to whether we've got the world right. This may be the Western Hemisphere's distinctive form of original sin, committed when Columbus mistook his landfall for India. The indifference is somewhat understandable. The British inhabited a tiny, vulnerable island, and so looked outward and produced the world's most doughty and observant travelers: people like Sir Richard Burton and Rebecca West. Americans have had an entire continent for distraction, and two oceans for insulation. The consequences are hard to shake. Walter Lippmann would become a distinguished commentator on foreign affairs, but he began as a typical American naif, setting out for what he thought would be a sunny European jaunt in June of 1914, right after the assassination of Archduke Franz Ferdinand had set World War I into motion. "It was possible for an American in those days," Lippmann later wrote, "to be totally unconscious of the world he lived in." It still is. Lynne Cheney, a former chair of the National Endowment for the Humanities and the wife of Vice President Dick Cheney, wondered aloud less than a month after 9/11 about a *Washington Post* article suggesting that Americans were failing to learn what they needed to know about other cultures. She brushed the criticism aside impatiently, saying that if there was a failure, it was the failure of Americans to know enough about America.

For whatever reason, geographic innocence is an ingrained American trait, confirmed with tedious regularity. 2001: A study by the Asia Society finds that a quarter of all high school seniors cannot name the ocean that separates North America from Asia. 2002: A study by the National Geographic Society finds that more than three quarters of Americans aged eighteen to twenty-four can't locate Iran or Iraq on a map. 2003: A report by a blue-ribbon panel assails America's "stubborn monolingualism" and "ignorance of the world." 2006: Another National Geographic study finds that a third of all young Americans believe that you'd have to travel north, east, or west to get from Japan to Australia.

Only two out of ten of the young Americans surveyed owned a passport.

All of this is aggravated by a new development, fueled by security concerns: the severe restrictions on foreign students coming to the United States and the harassment by federal agents of foreign students already here. Nearly 600,000 foreign students attend American colleges and universities, but applications have slipped sharply in some recent years — by 28 percent in 2004 — and schools in other countries are competing for those ambitious students. One British university administrator explains, "International students say it's not worth queuing up for two days outside the U.S. consulate in whatever country they're in to get a visa when they can go to the U.K. so much more easily."

It may be that in a busy and sprawling nation of 300 million, with a large and nearly empty territory to command attention, a lack of interest in the outside world among ordinary people should not count as unusual. CBS News, which in all of 2004 devoted only three minutes to the genocide in Darfur, presumably knows its market. But what about those who style themselves the elite, especially those whose job it is to monitor threats from the beyond? "I once asked an American general in Vietnam if he had read anything about the French experience in Indochina," a veteran foreign correspondent recently wrote, "and he said there was no point because the French had lost and, therefore, had nothing to teach us." America invests enormous sums in intelligence gathering—the so-called "black budget" runs to upward of $40 billion annually—but somehow doesn't have much to show for it. The intelligence agencies have been criticized for their lack of attention to the threat from militant Islam, but the underlying problem was hardly new. Three years before 9/11 a former CIA officer with extensive experience in the Middle East recalled that not one of the Iran desk chiefs who served during his eight years of working on Iran could speak or read Persian. Not one of the Near East division chiefs could read or speak Arabic, Persian, or Turkish. He wrote:

Sterling exceptions aside, the average senior officer rose through the hierarchy without ever learning much about the language, culture, or politics of the countries in which he served. . . . At the Agency's espionage-training school ("The Farm") at Camp Peary, near Williamsburg, Virginia, instructors regularly told trainees that cultural distinctions did not matter, that an operation was an operation regardless of the target. Whether Arab, German, Turkish, Brazilian, Persian, Russian, Pakistani, or French, targets were (as Duane Clarridge, a Europe Division and counterterrorism-center chief baldly put it) "all the same."

What accounts for such an attitude toward the world—this strange mixture of studied ignorance, intense involvement, and instinctive withdrawal? Is it a form of the "moral barrier" that some say separated insiders and outsiders in Roman eyes—and which also may have constituted an "information barrier"? Is it a sense of superiority? There's certainly some of each of these things. Like the Romans, Americans as a people habitually derive foreign policy from values rather than interests—a propensity that even facts may not dislodge. "They carried with them," Susan Mattern writes of the Romans, "an ideology of the foreigner with the authority of literary tradition. This ideology affected how they perceived their neighbors even after firsthand observation." Place this alongside the worldview of the senior American official who in 2002 mocked what he called the "reality-based community," which believes that "solutions emerge" from the "judicious study of discernible reality." And like Rome — "It is your special gift" — America has a driving sense of mission. In his novel *The Quiet American*, Graham Greene skewers America's anti-communist activities in Southeast Asia. But he's clear about the motivations of his main character, Alden Pyle. "He was determined — I learned that very soon — to do good, not to any individual person but to a country, a continent, a world. Well, he was in his element now, with the whole universe to improve."

The idea of American exceptionalism extends powerfully

through the national psyche from the founding of the first colonies up to the more recent efforts to somehow force-feed democracy to the Middle East. It begins with the proposition that those who came to America were fleeing the oppressive feudal and religious structures of an ailing Europe. Here, on a vast and virgin continent, new structures took shape, based on sturdy ideas of freedom and equality; and after a relatively mild revolution those values were officially enshrined under a government whose power could be checked in important ways.

In his defining study, *The Liberal Tradition in America,* Louis Hartz identified two basic components of the American stance toward the world that grow out of this sense of exceptionalism. The first of these is a strong isolationist impulse: "the sense that America's very liberal joy lay in the escape from a decadent Old World that could only infect it with its own diseases." Thomas Paine was wary of the alliance with France that ultimately brought American victory in the War of Independence. In the late nineteenth century the Anti-Imperialist League tried to keep America from following the path of the European colonial powers. America rejected the League of Nations after World War I. The United Nations is viewed by many Americans with distaste if not contempt.

The second component, according to Hartz, is a messianic streak: "Embodying an absolute moral ethos, 'Americanism,' once it is driven onto the world stage by events, is inspired willy-nilly to reconstruct the very alien things it tries to avoid." The American Idea becomes a commodity for export, maybe the only item of domestic manufacture that can't be replaced by cheap foreign knock-offs. It shaped the occupation of Germany and Japan. It animated the battle of ideas during the Cold War—Jimmy Carter's human-rights campaign, Ronald Reagan's crusade against the "evil empire." The messianic impulse is so deeply rooted that it emerges not only in contexts like Iraq, which tempted some in Washington with the possibility of transforming the entire

Arab world, but even in a context that is completely devoid of humanity, like the moon. "We came in peace for all mankind" was the announcement that Americans engraved on a plaque for the moon's surface—a message that, so far as we are aware, has been read by no one since it was left behind in 1969.

One corollary of a messianic tendency is the desire for public acknowledgment, for gratitude. The Romans sought "symbolic deference" from enemies and supplicants; the tombstone of a first-century governor proclaims proudly that he had brought barbarian leaders "to the riverbank that he protected" in order that they should "adore Roman standards." In 2004, as America looked for a suitable Iraqi to head a provisional government, the president laid down a single criterion: "It's important to have someone who's willing to stand up and thank the American people for their sacrifice in liberating Iraq." In 2006, amid the occupation's continuing horrors, President Bush held a meeting with his war cabinet and outside experts and expressed the view, according to one participant, that "the Shia-led government needs to clearly and publicly express the same appreciation for United States efforts and sacrifices as they do in private."

A messianic outlook makes no sense if it's not possible for other people to be saved. The Romans never shed their cultural stereotypes, but they came around to the view that "Roman-ness" was potentially accessible to anyone who embraced Rome's values. Americans are contradictory in just the same way. We harbor a full panoply of stock images—of Arabs, of Africans, of Germans and French. We can produce stereotypes while on autopilot. Once, some years ago, a poll asked Americans for their opinion of a fictitious ethnic group it called the Wisians, and respondents gave them a low favorability ranking: 4.12 on a scale of 9 (above Iranians and Gypsies, below Greeks and Koreans). At the same time, Americans tend to see "others" as being more or less exactly like ourselves; it's the default presumption. Or at least other people *would* be like us if only cer-

tain cultural impediments and institutional restraints were removed — if only they had democracy, for instance; if only they
had free markets and iPods. Human nature, in other words, is
basically American. This may be a comforting sentiment, but it
can end up enabling just as much ignorance as arrogance or dis-
dain does. The "just like us" argument is especially insidious if
you get the "us" part wrong. One great American myth is that
other peoples ought to be able to solve their ethnic and sectarian
differences peaceably, because "we did" — which, as Benjamin
Schwarz has written, simply ignores the vast amount of ethnic
cleansing and cultural obliteration that the American settlement
entailed, and that afterward kept the melting pot from boiling
over. Lynne Cheney has a point when she encourages Americans
to learn more of their own history, although history like this
probably isn't what she has in mind.

"Sameness" is a delusion of globalism. There is the "Mc-
World" sameness of popular commerce and culture — guerrilla
fighters in Madonna T-shirts, Google spoken everywhere — and
then there is the sameness of the elites. The sociologist Richard Florida, taking issue with parts of Thomas Friedman's "the
world is flat" thesis, argues that in fact the modern world is
"spiky" — most of the wealth, creativity, and entrepreneurship on
the planet is contained in some fifty city regions, and it is over-
seen by a class of people who move easily among these *de facto*
city-states and often owe allegiance mostly to the class itself. The
world appears flat, in other words, when you're at the tree-can-
opy level. The Roman Empire, too, was an urbanized and spiky
place. The elite of Rome felt completely at home in the thin slice
at the social top of, say, Antioch or Athens, Alexandria or Nar-
bonne. But did they connect with, or understand, the millions
of "locals" in the far-flung provinces, or even in the Italian countryside? Not really. "There seems to have been a sharp cultural
cleavage between the upper classes," writes A.H.M. Jones, "who
had not only received a literary education in Latin and Greek
but probably spoke one or the other of these languages" — the

Business Class and NPR crowd—"and the mass of the people, who ... spoke in a different tongue," such as Celtic, Gothic, Coptic, or Punic—Rome's version of flyover people and the NASCAR nation.

Like those of imperial Rome, America's elites are an urban and international group, perhaps on their way to forming a distinct transnational class. They are cosmopolite-citizens who often have more in common with members of that same class around the world than with other members of their own society. The elites of Washington, New York, Los Angeles, and Boston may be American by birth, but their wall hangings are from Peru, their sculptures from Nunavut, their literary fiction from Sri Lanka, their CDs from Brazil, their basmati from India, their wine from New Zealand. Their religious values, if they have any, may be drawn impressionistically from Eastern and Western traditions—an eclectic pantheon.

"The empire was ruled by an aristocracy of amazingly uniform culture, taste, and language," writes the historian Peter L. Brown.

> In the West, the senatorial class had remained a tenacious and absorptive elite that dominated Italy, Africa, the Midi of France, and the valleys of the Ebro and the Guadalquivir; in the East all culture and all local power had remained concentrated in the hands of the proud oligarchies of the Greek cities. Throughout the Greek world no difference in vocabulary or pronunciation would betray the birthplace of any well-educated speaker. In the West, bilingual aristocrats passed unselfconsciously from Latin to Greek; an African landowner, for instance, found himself quite at home in a literary salon of well-to-do Greeks at Smyrna.

In modern terms this would be DavosLand or AspenWorld, and if it's where you happen to live, you can be misled into thinking that it's a very large place, and the only one on the planet that matters.

"All I Need to Know . . ."

THE AMERICAN-CONTROLLED Green Zone in Baghdad serves as a microcosm of American attitudes toward the non-American world: the innocence, the optimism, the arrogance, the ignorance, the idealism, the zeal. When American forces captured Iraq's capital, in April of 2003, they created a headquarters zone in the four-square-mile precinct of palaces and parks, hotels and villas, military monuments and government buildings, that lies at the very heart of Baghdad. As the security situation began to crumble, a few months after the occupation began, the Green Zone was sealed off by concrete blast walls ten feet high. The U.S. administration took up its work under the blue dome of Saddam Hussein's former Republican Palace; the Green Zone was soon home to thousands of civilian and military personnel. L. Paul Bremer, an American diplomat, was brought in to administer "free Iraq" from inside this Praetorium, and despite his official title ("administrator of the Coalition Provisional Authority"), he was routinely referred to in the press as the American "proconsul," a term that harked back two millennia to Rome.

The unstated premise of the Green Zone seemed to be that Iraq was a *tabula rasa* on which Washington's operating procedures — its "civilizing mission," as Velleius Paterculus might have put it — could easily be inscribed. Prior to taking up his post in Iraq, the proconsul had pursued a typical *cursus honorum*, American style: Ivy League education, foreign service, ambassadorship, Kissinger Associates, Marsh & McClennan, farmhouse in Vermont. He had never been to Iraq. Behind the Green Zone's walls the degree of isolation from Iraq was felt by everyone but gave pause to few. Virtually no one spoke any Arabic (and translators, scarce at the outset, would be in short supply even for front-line fighting units three years into the war). "From inside the palace," one American adviser later wrote, "staff members often had to call home to the United States to ask family and friends about what had just happened only a mile away by means of mortar, rocket, or car bombings near the Green Zone." (This adviser, who had brought a George W. Bush "Mission Accomplished" action figure with him to Iraq, finally became disillusioned.)

Taking an image from mathematics, the historian Charles Maier argues that empires tend to be organized according to "a fractal set of hierarchies"—that is, the structural principles (and flaws) of the whole are replicated at every level in those of the parts, and vice versa. The Green Zone is a fractal domain. Bureaucrats and civilian experts representing scores of government agencies, oblivious of culture or history, were brought in to create an embryonic version of American government for the Iraqis to adopt as their own. Americans were enlisted to help draft a new constitution. They drew up scores of new American-inspired laws to address even the least urgent matters, such as patents and copyrights and other kinds of intellectual property. They created shadow ministries of agriculture, education, electricity, human rights, oil, trade, youth and sports, and more. Monday-night seminars were held to teach prominent Iraqis the basics of a free-market economy. ("I hoped that these sessions would evolve into a sort of Council of Economic Advisers to the interim government we intended to set up," Bremer recalled in a memoir.) One department, seized with a desire for tax reform along neoconservative lines, sought to put into place a 15 percent flat tax—this in a nation where taxes had been neither levied nor paid. Another official devoted his tour of duty to implementing a traffic code for the entire nation of Iraq, taking as his model the traffic code of the state of Maryland.

One Iraqi employed in the Green Zone—as, among other things, a translator for Paul Bremer—recalled an incident involving an American military attaché:

Lt. Col. Bill came over to my office to enlist my help in finding him an interpreter. He complained that the officials of the former Ministry of Culture could not speak English. Pointing at a book on my desk, he asked, "So what is this book you are reading?"

"The bible," I started to say.

"But I thought you are Shia!" he cut me off.

I told him that the book, *The Old Social Classes and the Revolutionary Movements of Iraq*, was Hanna Batatu's book on Iraq, and that it was a must-read for anyone dealing with modern Iraq's political history.

Bill pulled from his back pocket a green paperback, published in the mid-'80s by Iraq's Ministry of Tourism, and with a straight face told me, "All I need to know about Iraq is in here."

This kind of outlook explains a fiasco in Somalia in 2006. With few reliable indigenous Somali sources and little cultural understanding, but quick to impose a preconceived template on an evolving reality, American intelligence officers misinterpreted a series of events on the ground as a terrorist resurgence, and in response began funding local warlords to serve as a counterforce. Heavily armed and flush with cash, the warlords themselves became a source of trouble, inviting a powerful local backlash and the rise to power of a militant Islamist with reputed ties to al-Qaeda — the exact reverse of the intended outcome.

Unlike Denmark or Costa Rica or most other countries, America has an impact on every other place in the world. Even without "unapologetic and implacable demonstrations of will," America's behavior influences the price of gasoline in Mexico, the duration of airport delays in Singapore, the television lineup in Jordan, the standard of living in Shanghai, the spread of AIDS in Botswana. America's impact may be unprecedented in its scope, but the phenomenon itself is one that Rome knew well. And so is the phenomenon of blowback. Everything America touches can potentially touch us back — often unpredictably, and maybe not for years. In its proxy war against the Soviets in Afghanistan, America trained and armed Muslim warriors from all over the world, many of whom turned into the multinational jihadists we are fighting now. As part of the war on drugs, Washington encouraged farmers in South America to stop growing coca and to start growing flowers — creating a big export crop and undermining America's own flower industry. For more than half a century American troops have helped to safeguard South Korea's security, giving rise to positive Korean words like *yonmi* ("associate with America") and *sungmi* ("worship America") — but also to words of reaction like *hangmi* ("resist America") and *hyommi* ("loathe America"). Nothing is more symbolic of America's global influence than the Internet, developed initially by the defense industry: even the electrons do America's bidding! But this same globalizing force enhances the economic power of other nations, and puts insidious new tools into the hands of those who would do America harm. The blowback problem is a reality that can't be eliminated. But ignorance makes the problem worse, by concealing the fact that it's even there.

On the Fourth of July in that first year of the American occupation the explosions over the Green Zone were not mortar rounds but fireworks, like the ones on the Mall in Washington. In so many ways the enclave was a replication not just of America but of the American capital and its mindset, as if this form of transplant would cure what ailed Iraq. Green Zone telephones were even given American area codes — 703, one of the area codes for metropolitan Washington, which happens also to be used by the Pentagon; and 914, which designates Westchester County, in suburban New York. If all the world is potentially an America in embryo, then why shouldn't the Green Zone be a local call?

5

THE BORDERS
Where the Present Meets the Future

The barbarians were adapting themselves to Roman ways, were becoming accustomed to hold markets, and were meeting in peaceful assemblages. They had not, however, forgotten their ancestral habits, their native manners, their old life of independence. . . . They were becoming different without knowing it.

— Cassius Dio, *Roman History*

Ai pledch aliyens to di fleg
Of di Yunaited Esteits of America.

— phonetic guide handed out at an immigration rally, 2006

HADRIAN'S WALL, the great stonework fortification that stretches across the neck of northern England like a thrall's collar, occupies a romantic place in the historical imagination. My own introduction came one long-ago fall morning when, as a student, I cracked open a new third-year Latin textbook and came across a color photograph of the wall in winter, dusted with snow, relentlessly climbing up crags and down defiles, until it disappeared into a horizon beyond time. The wall, I read, had been built at the command of the emperor Hadrian, beginning around 120 A.D., to delineate the boundary between Roman Britain and the barbarian hinterland to the north, and it stretched seventy-five miles from the mouth of the Tyne, on the east, to the Solway Firth, on the west. There was a gateway in the wall at every mile, and a pair of turrets between each two milecastles. Every manned point was visible to two others. With torches, I imagined, you could relay a signal from the North Sea to the Irish Sea in minutes.

In one of those reinforcing coincidences that can fix a memory for all time, soon after seeing that textbook photograph I came across Rudyard Kipling's *Puck of Pook's Hill,* which contains this evocation of what it might have been like for a Roman soldier arriving at the northern British frontier for the first time:

> The hard road goes on and on — and the wind sings through your helmet plume — past altars to legions and generals forgotten. . . . Just when you think you are at the world's end, you see a smoke from east to west as far as the eye can turn, and then, under it, also as far as the eye can stretch . . . one long, low, rising and falling, and hiding and showing line of towers. And that is the wall!

Reading those words even now I feel a rush of excitement, and involuntarily pull an imaginary cloak a little tighter around my shoulders.

To be sure, in the course of fifteen centuries the wall has been dismantled in many places, as people used its expertly cut building stones for houses and sheep pens and churches nearby. Hadrian's Wall was the original Home Depot. Even in the best-preserved stretches the wall is now only about six feet high (and about six feet thick), rather than the fifteen feet high of Hadrian's time. But no matter; it is an impressive sight. I've walked long stretches of the wall on two occasions in the past few years. My first view of it came on a ridge a little north of the old Roman camp at Vindolanda. A bend in the road, and there it was: a serpentine bulwark running up and down along the beetle-browed ridges. It was wintertime, close to the solstice, and that far north the sun sets very early—around 3:00 P.M. But in truth the winter sunset heralds two hours of dusky purple and orange in the southwestern sky, above a serrated landscape visible to a range of thirty miles. The low sun brought into sharp relief the system of defensive trenches and earthworks that parallel the wall on both sides. Each step left a trace in the frozen grass. To a Roman soldier looking south, the first olive tree would have seemed very far away indeed.

Northumbria was a Roman military zone, and leaving aside the wall itself, signs of the Roman presence are everywhere. A major Roman road, the Stanegate, runs east to west, still as straight as a rule in many places, though mostly a country track and sometimes a farmhouse driveway. One glance at a 1:25,000 ordnance map reveals Roman remains, some excavated and some not, every half inch or so. Here and there the words "cultivation terraces" and "Roman aqueduct (course of)" suggest how the legions sustained themselves. The Roman relics are interspersed on the map with designations like "tumulus" and "cairn," remnants of the culture that the Romans came among but never eradicated.

The Romans arrived in Britain with Julius Caesar in 55 B.C., though the arrival constituted little more than what nowadays would be called a toe-touch. They arranged some treaties with local magnates, took some precautionary hostages, and then left, not returning until a century later, in 43 A.D., during the reign of Claudius. This time the Romans pushed north, and eventually, by the late 80s, had advanced well into Scotland. They built strings of camps and forts right up to the edge of the Highlands before falling back in the second century and defining a frontier farther south with the construction of Hadrian's Wall. When artists conceive the building of the wall, they tend to portray a Roman legionary with a bullwhip supervising a chain gang of fur-clad native laborers—which is not the way it was done. The wall was designed, surveyed, and constructed by the Roman soldiers themselves. They did their own heavy lifting. One nineteenth-century calculation estimates that building the wall would have taken 10,000 men about two years to complete.

It's odd and a little sad to think of this work of military engineering as dividing Romans and Scots, two of the greatest engineering peoples of all time. Or so it seems to do: Hadrian's Wall has the appearance of something built to repel the barbarian hordes. Like a pinhole in a space suit, you think, any breach would spell disaster. That impression is misleading. Hadrian's Wall was not meant to be a Maginot Line or a Berlin Wall. It was built to mark a frontier, but it was also meant to be penetrated. Of course, its sheer size would have been awe-inspiring to the people beyond — in an era without aerial surveillance, the cold face of a high wall would be a forbidding sight — and it certainly would have deterred significant cross-border attacks. But the milecastles had fortified gateways expressly to make the wall permeable — to regulate cross-border traffic rather than to prevent it. Commerce moved in both directions, and Roman soldiers and traders were active in the territory to the north. At major crossing points in the wall, towns that would be home to both Roman newcomers and indigenous Britons — those *Brittunculi* — grew up symbiotically outside the military installations. You would have seen the same pattern, but more intensive, around the big bases on the Rhine and the Danube, like Cologne and Mainz and Regensburg.

And you can see the pattern today in the cities and towns along the length of the U.S.-Mexico border, fed by human movement and international trade — although relative to Rome the scale is immense. The Roman Empire was not thickly settled; at its height the total population was no more than that of modern France, perhaps 50 or 60 million and skewed toward the East. By itself the urban agglomeration of modern El Paso–Juarez is three times the size of the ancient city of Rome. But you'll find in El Paso, on the American side, the same imperial tension between separation and integration, between sepsis and symbiosis. The border, at that point a thin line of polluted river spanned by ugly modern bridges, is both a gash and a suture. Car and foot traffic at the several ports of entry is incessant. One way or another, Juarez, on the Mexican side, lives off America. America provides a market for the cocaine and marijuana going north. America is the source of the used cars and appliances heading south, and sold in *yonke* shops everywhere. More important, America operates the hundreds of maquiladoras in Juarez, assembly plants whose half million badly paid workers inhabit the city's squalid *colonia*. In the center of Juarez the main commercial strips are lit by neon signs for dentists and optometrists, drawing American patients who find health care in their own country too expensive.

The El Paso side of the border is home to Fort Bliss, an Army training center the size of Rhode Island, and once the staging ground for the campaign against Pancho Villa, in which my grandfather served as an aviator. The so-called Punitive Expedition of 1916–1917 was the kind of cross-border operation that a Roman commander like Germanicus would have understood (though it did not earn for General John J. Pershing, one of its leaders, the sobriquet "Mexicanus"). Fort Bliss is now a major deployment center for troops heading out to the wars in Iraq and Afghanistan, wars in the service of a far different notion of border defense. At the airport I noted some of the names on the desert cammies of soldiers in transit: Guttierez, Herrera, Corona. Many retired military personnel have settled around El Paso, forming *coloniae* of the Roman kind — veterans' settlements, not squatter camps, with the aging legionaries entitled to *praemia* like cheap food and medical care. El Paso is also the site of the National Border Patrol Museum, whose gift shop sells tie tacks in the shape of handcuffs, and a book on failed desert crossings by illegal aliens, titled *Dead in Their Tracks*. The city is home to the El Paso Intelligence Center, which is intended to serve as America's never-dormant brain in the fight against illegal drugs and immigration. From its base in El Paso the Drug Enforcement Administration sends frequent raiding parties into Juarez, at Mexico's invitation. Unlike Julius Caesar, they do not burn down the bridges after them when they return.

On the U.S.-Mexico border the long reach of Washington is always visible and in some ways formidable. And at the same time it seems shaky and tenuous, and strangely foreign.

Where Do "We" Stop?

THE NOTION OF A BORDER, like the thing itself, is meant to be a clarifying concept, at least in someone's eyes. "A border," says John Sayles, "is where you draw a line and say, 'This is where I end and somebody else begins.'" But invoking the word often makes things less clear, more complex, harder to understand. What is meant by "border" anyway? Depending on the circumstances, it can refer to a political jurisdiction, or an economic boundary, or an ethnic or cultural or religious divide, or a psychological state — or (rarely) all of these things neatly conjoined, as in Eamon De Valera's deluded vision for Ireland, or any fervent nationalist's dream of Greater Blankistan. A border can be seen as temporary and fluid or fixed and immutable. Some things, like weather and climate, don't respect borders at all. Others, like plants and animals and pollution and disease, mostly don't respect borders, though sometimes, with a lot of effort, they can be encouraged to. Certain kinds of borders demarcate not the exterior but the interior, like the boundaries of class that the ancients knew so well, and that Americans are fast erecting.

To say that the word "border" has no single fixed meaning or reality is not to say that borders are meaningless or unreal — only that they are complicated. Rome maintained its frontier during a time before global communications and electronic capital flows brought the idea of borders into the fourth, fifth, and nth dimensions; even so, Rome's borders remain a subject of intense disagreement. America needs to concern itself with many kinds of borders — and incursions across them — that are totally new. Hadrian's Wall would today have to be supplemented by Hadrian's Firewall. At the moment, DNA marks a border for each of us that is mainly personal; but as biometrics are used to register identity and thus to define citizenship, it could someday mark a political border as well. DNA testing is already being used in immigration cases. Other borders are disappearing altogether: consider the transnational revolution represented by that ubiquitous street-corner shrine, the ATM machine. A global attack on computer encrypting systems, or the onset of mutations caused by genetic experiments, or the disappearance of fresh water from lakes and rivers, or the arrival in one place of industrial pollution generated somewhere else: these are all border issues in their way — border issues that Rome could not have conceived of. They are as disruptive and dangerous as any invasion by barbarian marauders.

But Rome also faced some of the very same border issues America does. Rome's economy may have been primitive, but for its time the empire was already a "globalized" place. Rome was no stranger to dislocations caused by worldwide market forces, or to the riotous and unpredictable interplay of ethnicities and cultures and religions. And when it comes to the physical aspects of borders — real human beings, and the real ways they make a living, and the real geography they inhabit and control (or would like to) — the dynamics haven't changed fundamentally in two millennia. But there's something of a surprise here. In popular shorthand the long saga of Rome and the barbarians is typically held up as a case study in failure. That turns out to be a narrow view — not entirely true, and for an American, not a helpful perspective at all.

The Roman imperial land boundary in a particular place was known as the *limes* — the plural is *limites,* as in "limits." During the past half century experts have been gathering regularly for a Roman *limes* conference (or, in Teutonic scholarly idiom, a *Limes-kongresse*) to thrash things out. These gatherings were started by the great archaeologist and historian Eric Birley, the father of Vindolanda's Robin Birley. One scholar may reveal new archaeological findings based on aerial surveys of southern Algeria. Another may invoke Frederick Jackson Turner, looking at patterns of war and accommodation as Americans pushed their shifting frontier across a continent. Yet another may draw comparisons with ancient China, which had border problems of its own in the north (and put up another well-known wall).

The Victorian historian J. R. Seeley memorably declared that the British had acquired their empire "in a fit of absence of mind" — that is, through an unpremeditated cascade of opportunities and impulses, rather than some prior grand vision of what an empire should look like. Maybe all empires begin like that. The Romans, in any event, acquired their empire in much the same way, accruing territory in chunks large and small, sometimes for obvious strategic reasons, sometimes in response to the tactical urgencies of the moment, sometimes because of nothing more than a general's or an emperor's bid for acclaim. Glory was no small motive in imperial Rome. (It's no small motive now, though in America it tends to dress in the plainer fabric of "legacy.") According to Tacitus, the emperor Tiberius once urged his stepson, Germanicus, who was busy waging war in Germany, to "leave his brother . . . some chance of distinction"; the emperor's biological son, Drusus, would be needing to earn triumphal laurels of his own, and "in the absence of enemies elsewhere" Germany offered the only opportunity.

Whatever the motivations, by the middle of the second century A.D. the territorial form of the empire had achieved more or less its ripest fruition. Some boundaries — the ones marked by the ocean — were obvious and unyielding. Others were boundaries of choice, but were suggested and demarcated by geographic

features like the Rhine and the Danube. These rivers could be
bridged at will by the Romans—Trajan, on campaign, once built
an arched span on twenty stone piers in the swift waters below
the Iron Gates of the Danube—but crossed by anyone with a
boat or when they froze. Still other boundaries, the ones that
ran across formless tracts of land, heedless of natural barriers,
and perhaps right through the domains of whole peoples and
cultures, were simply decisions—decisions based on a variety
of factors, and decisions that could be changed. Hadrian's Wall
didn't have to be planted precisely where it was (and for a while
the Romans maintained a different frontier line, the earthen Antonine Wall, a hundred miles to the north). The specific path of
the line of ditches, ramparts, and forts that ran for 340 miles,
roughly between Bonn and Regensburg, and plugged the gap between the Rhine and the Danube was not preordained; for one
fifty-mile stretch the *limes* runs straight as a plumb line, up hills
and down ravines and across rivers, as if the real point were to
say to non-Roman peoples, "Think twice before you meddle with
us, the Makers of Straight Lines!" In the deserts of North Africa
the long boundaries, sometimes indicated by walls and trenches
in the middle of nowhere, like the *fossatum* in Algeria and Tunisia, represent sheer acts of will. The Euphrates may have made
sense as a border, and was sometimes used as such, but just as
often it served as a highway beckoning to the rich lands beyond,
as it had beckoned Alexander the Great. On numerous occasions
proud Roman armies sallied across those parched Middle Eastern plains and into Mesopotamia; often, much smaller armies
hobbled back. The target was generally Ctesiphon, the capital of
the Parthian and Persian Empires, a few leagues south of modern Baghdad. Military transport planes supplying America's own
Mesopotamian war rattle Ctesiphon's ruins every day as they fly
into Rasheed Air Base.

What did borders represent in the Roman mind? The answer
from historians fifty or a hundred years ago might have invoked

the grim specter of *barbaricum,* pure and simple, its warriors bat-
tering at the gates from every side, like Tolkien's hideous orcs.
The authors of the *Cambridge Ancient History,* written in the
1920s and 1930s, take this point of view: "All along the borders
of the civilized world there stretched a belt of turbulent peoples who were ignorant of the restraining influence of civilization but were eager to gain for themselves the riches it had produced." A few decades later, in the aftermath of World War II,
as Cold War thinking settled heavily upon the West, another authority referred to the line between Rome and the barbarians as
an "iron curtain" and spoke of the Rhine and Danube frontiers
as representing a "moral barrier" beyond which (in Roman eyes)
lived pestilential and scarcely human peoples. In dealings with
them, the customary rules of civilized behavior hardly applied:
"There was no Roman historian who did not narrate with com-
plaisance how their armies burnt down the settlements and dev-
astated the cornfields of the barbarians; they report wholesale
merciless massacres without showing any moral scruples. And
more than that. The Romans always had the intention — often
avowedly — of completely eradicating entire populations." Evidence of this outlook is not hard to find. The emperor Domitian, after wiping out one tribe in North Africa, observed coolly
that he had "forbidden the Nasamones to exist." Moral barrier
or no, a world of difference separated Roman civilization and
the cultures of the outside tribes.

But antagonism doesn't define the whole relationship. Nor do
the words "fixed" and "inviolate" capture the Roman notion of
frontiers. Americans today are accustomed to thinking of physical borders as a static sort of artifact — not as unchanging as, say,
the path of the equator, but more durable than the outlines of
a congressional district in Texas. It's because they're durable —
sacred — that the General Services Administration has enlisted
prominent architects to build three dozen new border-station
gateways, at a cost of up to $100 million apiece (and sent the de-

signs to tour the country, in an exhibit called "Thresholds Along the Frontier"). In part this American view of borders reflects the limits of a blinkered human lifespan: change typically presents itself in nearly undetectable increments. It also reflects the relatively easy experience of America, once the transcontinental dictates of Manifest Destiny had been satisfied. Elsewhere in the world, territorial flux has in fact been the norm, and has been extreme. Half the world's national boundaries are less than a century old. At any given moment three or four dozen boundary issues are in various states of violent or diplomatic dispute. Americans look at what we have and regard it as not only foreordained but almost "normal."

In its most simplistic form the debate over how Rome saw its borders breaks down into two camps, divided over the answer to the question, Why did the frontiers stop where they did? The more traditional camp holds that after centuries of expansion, when the notion of setting fixed borders was on no one's mind, the Romans at last made a deliberate strategic choice to set limits in a specific way. At some point after the Teutoburg Forest disaster, Augustus wrote out instructions for his successor, in his own hand, it is reported—a kind of final testament, counseling Tiberius against further imperial adventures and "advising the restriction of the empire within its present frontiers." Let's not get greedy, he seems to have said; let's consolidate and fortify and hold the perimeter—that job is going to be tough enough. One historian has called this Augustan injunction "the view of a weary and frightened old man whose fingers had been badly burned." To American ears the warning recalls the farewell addresses of both George Washington (who cited the perils of "foreign entanglements") and Dwight D. Eisenhower (who cited the perils of a "military-industrial complex"). It also resonates with the American view of what national borders should be: clear and fixed.

A newer camp, represented by historians like C. R. Whittaker and Benjamin Isaac, agrees that defending some sort of perimeter was important, obviously, but argues that the Romans had no fixed notion of where that perimeter should lie. And besides, their whole outlook would have resisted the idea of explicit limits: these are the "empire without end" people. Stone monuments have been found that marked the border between Roman provinces, and between districts inside the city of Rome, Whittaker observes, but none that announced the edge of the imperium itself, the equivalent of "You are now entering the Roman Empire. Please pillage safely." There was no Pritzker Prize or traveling museum show for high-concept Roman border stations. In this newer camp's view, borders were established in different places for different reasons. Sometimes the explanation might be ecological or logistical: any farther, and you couldn't support or supply an army. Sometimes it might be political, hinging on events in Rome or diplomacy abroad. Whatever the case, fortifying a frontier — even building a wall — needn't mean that Rome's expansionist ambitions were at an end. There were Roman forts and settlements, and penetrating roads, north of Hadrian's Wall, east of the Rhine, north of the Danube, and south of the African *fossatum*. As for the advice of Augustus: it is eloquent and instructive, and perhaps wise, but did it really set Rome's course? Americans, after all, cite the warnings of Washington and Eisenhower with reverence, but honor them in the breach. Plenty of emperors after Augustus indulged in foreign adventurism, not behaving "defensively" in the least.

Maybe the Romans wanted to have it both ways, as Americans do. America isn't grasping for more territory; our substitute for territory — our version of *imperium sine fine* — is the ideology of democracy and free markets. Should we push them aggressively, even by means of pre-emptive war, hastening the world toward its destiny? Should we hold back, to conserve the tenuous gains we've made and let the impetus of history gradually have its way? Well, we don't do either with any consistency.

Mark Twain, anti-imperialist to the core, mocked America's vacillation: *"Shall we?* That is, shall we go on conferring our Civilization upon the peoples that sit in darkness, or shall we give those poor things a rest?"

Historians may clash in the debate over Rome's borders, but almost everyone now sees the imperial frontier not as a solid membrane but as a dynamic zone where the interactions of different cultures had transformative repercussions in a wide band on either side. To Americans, this is a very recognizable picture. It's a lot like the picture that shows Mexican workers canning fish in Alaska, and retired Americans driving hundreds of miles to get false teeth in Juarez. It's the picture Americans see as they watch manufacturing jobs heading south and drugs heading north. It's the one that shows sales of salsa finally overtaking sales of ketchup, not just along our southern rim but nationwide. It's a picture of the future unfolding before our eyes.

The Onslaught That Wasn't

TRAVEL THE 6,000-MILE CIRCUIT of the Roman Empire, and in the borderlands beyond you would see hundreds of tribes that viewed themselves as distinct, even if the Romans, looking from inside out, lumped many of them together into a barbarian mass. But peering into the empire from anywhere on the outside, you really *would* see something remarkably uniform wherever you looked. This was the case even though the empire encompassed a wide variety of peoples, not all of whom knew their butter knives from their fish knives before coming under Roman rule. The empire's cities, sometimes laid out by its legions, often had similar street plans. The fields, plotted by Roman surveyors, were of a recognizable size and shape. The roads were standard, and the very act of building one called new towns into existence, just as America's railroads and interstates would one day do. The army was fitted out in standard fashion with standard equipment, whether in Jerusalem or Córdoba, Lepcis or Londinium; the legionaries all wore something akin to dog tags, like American GIs — and so did their horses. Even the statuary from place to place, thanks to ingenious techniques for replicating marble sculpture, was often exactly the same. At one time, it has been estimated, there were some 20,000 statues of Caesar Augustus publicly displayed around the empire, making the post-office photographs of American presidents seem pathetically unambitious by comparison. All of this was just the physical embodiment of a powerful underlying organizational and social dynamic, which made Rome the most irresistible force for assimilation the world had ever known.

The frontier was a crucible. Two cultures, Roman and Other, were drawn together by trade: pottery, jewelry, salt, iron, and cloth going one way, from inside to outside; and cattle, horses, hides, food, slaves, and paid labor going the other way. Roman traders were active throughout *barbaricum*. Tacitus tells how, after a coup dislodged one barbarian chieftain, "traders out of the Roman provinces" were discovered cowering in his palace. (You'd find a lot of American businessmen in the same position after a coup in Kazakhstan or Georgia, or if the Green Zone in Baghdad ever fell.) Sometimes the people outside the empire had specialty items on offer. The tribes of the Caucasus were renowned for their stout, swift horses — exactly what the Romans needed when they developed heavily armored cavalry, modeled on that of the Parthians. (The frozen remains of sixty-nine ancient horses were discovered in the ice in the Altai Mountains a few decades ago, revealing the powerful legs and strongly developed shoulders that made them so desirable.) Map the places where Roman artifacts — jewelry, money, roof tiles — have been found in barbarian lands, and it's like watching an injected dye highlight a vascular system: the dots cluster along the river valleys and ancient trading routes that lead away from the empire's edge, strung out for fifty or a hundred miles and more, and then gradually thinning. German gets its word for "to buy," *kaufen*, from the Latin *cauponor*, meaning "to trade," and its many other borrowings illustrate the Roman contribution to tribal development: for instance, the words for "window" and "chimney" and "coin." (Contributions in the other direction are not so vast; the only word left by the Visigoths to the Latinate language of Spain is the word for "executioner.") The barbarians were fast learners. They adopted Roman farming and breeding techniques; you can tell from bones excavated beyond the *limes* that their cattle started getting larger, like Roman cattle. The impression that many people have of the Huns is of a frenetic race who lived in their saddles and suffered from a very hostile form of attention deficit disorder. But although personally uncouth, and an eater only of meat, Attila employed several Latin secretaries to conduct his correspondence. The one eyewitness account we have of a visit to Attila's capital, its location now lost, describes the "baths" he had built there, with imported stone, in imitation of a typical Roman facility. (The Roman hostage who built it, thinking this would be his ticket home, was instead made to stay and run the establishment — shades of the unfortunate Tony Last in *A Handful of Dust*.) During times of war, Rome might clamp down on trade in strategic items like iron and whetstones (and profiteers would of course find ways to elude these export controls, as corporations do today), but for the most part people were free to come and go, and to trade, without notice or interference. In many places the Romans would try to channel commerce, by means of bridges and walls and dikes, and they were ever on the alert for movements of people en masse. But there weren't tight border controls as we think of them. There weren't passports.

The word "barbarians" is still used by modern historians writing about Rome — political correctness has a limited amount of retroactive influence — but more often you'll see them using the names of particular groups; a few of them, like Whittaker, use a term familiar to Americans: "immigrants." In this Roman context the word jumps out, conditioned as we are by historical perspectives two millennia in the making. The Sack of Rome by the *Immigrants?* And the word imperfectly describes the outsiders who continually moved into the empire, in freedom or in some form of servitude, as individuals or in groups, in the open or by infiltration, in peace or through force. Still, in the course of the fourth century alone, perhaps a million people came into the empire this way, through countless "undramatic adjustments," as the historian Walter Goffart describes it, creating broad swaths of somewhat Romanized barbarians and somewhat barbarized Romans. By and large the barbarians came not to destroy what Rome had to offer but to get some of it for themselves, in the form of land, employment, power, status. First in the borderlands of empire and then farther inside, they were given jobs that Romans didn't want or couldn't fill, in the fields and the mines and the forts, including jobs as seasonal laborers. Then as now, the military was a transformative institution, the most important one of all, taking in rough-hewn raw material and turning out something a little more finished: the sort of person who might note on an epitaph that he was born a Frank and died a Roman centurion. It has been said of Kipling that the Roman soldiers in his stories resemble British subalterns in the Indian army. A bootstrapping Germanic graduate of a Latin as a Second Language program would have been more nearly the Roman norm. As time went on, the Roman military produced its own antecedent versions, many times over, of America's General Ricardo Sanchez (the son of Mexican immigrants) and General Colin Powell (the son of Jamaican immigrants) and General Peter Pace, the current chairman of the Joint Chiefs of Staff (the son of Italian immigrants).

The more elite among the non-Romans — the aristocrats — passed into the Roman hierarchy with relative ease. You see it first, in the early days, with non-Italian men from conquered

regions—from Gaul and Spain and Africa—showing up in the
senatorial ranks in Rome. (They were greeted by old Romans
with skepticism or snobbery, to be sure.) Arminius will always
be remembered as the duplicitous architect of the Teutoburg
Forest disaster, but his family also exemplifies how quickly barbarian nobility was absorbed into the Roman apparatus. His
brother Flavus served in the Roman army, even after Teutoburg. His nephew Italicus actually lived in Rome. His brother-
in-law Segestes remained with his tribe but was a Roman ally.
A son of Segestes became a Roman priest in Cologne. The late-
Roman period teems with prominent men of barbarian stock
who capably supported the empire. Think of Stilicho, the son
of a Vandal, who commanded armies for the emperor Theodosius and later served, in effect, as the regent for the emperor's
young son, Honorius. It was Stilicho who fought off or placated
Alaric and the Visigoths, year after year. And it was only after
palace intrigue brought about Stilicho's downfall that the way
was opened for Alaric's devastating progress through Italy. Alaric
himself had once served Rome as a commander of *foederati,* and
no small part of his complaint against the Romans stemmed
from their failure to promote him to a position he thought he
deserved—one of real consequence in the imperial hierarchy,
and with a Latin title to accompany it.

The image most of us have of the various barbarian invasions probably comes from maps in textbooks showing the Roman Empire puffed out to its fullest, with menacing black arrows labeled "Ostrogoths," "Visigoths," "Vandals," "Franks,"
"Saxons," and "Huns" arcing out of central Europe and punc-
turing the rest of the continent, a few lodging in Britain and Africa. The arrows are usually scored with little dates—378, 410,
455—showing the relentless advance. The maps bear a resem-
blance to diagrams of football plays; or, more striking, to maps
of Wehrmacht operations in World War II—replace "Attila" and
"Fritigern," say, with "Rommel" and "Paulus." Inevitably they

convey the impression, as historians caution, that the invasions were happening all at once — were maybe even coordinated — or at least that they occurred in the course of some compressed period of time. The maps also convey a sense of unstoppable power: a swarm of ferocious barbarians racing across the northern European plain, followed by lights out. The imprint of such maps on common knowledge and popular culture is probably indelible. In a long reflection about the 1992 Los Angeles riots, which had exposed the city's deep racial and ethnic animosities, the writer Jack Miles recalled a symptomatic conversation: "On the day after the first night of the riot, one of my colleagues said to me, as we left to hunt for a still-open restaurant, 'When the barbarians sacked Rome in 410, the Romans thought it was the end of civilization. You smile — but what followed was the Dark Ages.'"

The reality was more complicated. The invasions that occurred were costly in lives and treasure, and a threat to Rome's political control. But they played out over a period of centuries, usually with time to recover in between; and in each case they affected specific places rather than the whole empire. Some cities were overrun and ravaged, but not as many as you might think. The documentary sources are often so unreliable that it's not always clear when (or if) certain incursions even happened, or how significant they were if they did. The historian John Drinkwater maintains that emperors were not above picking fights and "demonizing the local barbarians" in order to prove their resolve, validate their regimes, and demonstrate that they were "doing a worthwhile job," thereby solidifying a place in the history books. (This is hardly an obsolete dynamic: Theodore Roosevelt and Bill Clinton, as one commentator observes, "both wished out loud that they had had the chance to show their mettle during a national crisis on the level of a major war.") The first round of invasions, in the third century, was eventually repulsed, through enormous military effort; the empire's borders were restored and even slightly expanded. The second round, beginning late in the fourth century, would in the end be handled differently, with some barbarian groups permitted to settle permanently in specific territories under their own leaders. Roman chroniclers throw around some large numbers — "carelessly transmitted numerals," C. R. Whittaker dryly observes — about the size of the invading armies, but the number of barbarians involved in each of these episodes was relatively modest; Whittaker cautions against the traditional rhetoric of "waves, floods, funeral pyres, and boiling pots," or the image of "nations on the move." Walter Goffart writes, "It would probably be wrong to estimate the size of any tribe in more than five figures, a trivial number in comparison to the millions of Roman provincials." And that's for the invading population as a whole; the actual fighting force in each instance might be no more than 5,000 to 15,000, and sometimes was fewer. Whittaker goes on to caution that the invasions should be seen not as if they were common-source disease vectors, single-minded and unremitting, but, rather, as individual events with individual contexts. Sometimes the barbarians were officially invited inside the empire; sometimes they were enlisted as allies by rebellious Roman generals; sometimes, as in Britain, the invaders came only after the Romans had left. The barbarians tended to be fragmented and fractious, their coalitions often falling apart in short order. If they acquired power, they typically wielded it as Romans would, through existing local institutions. They needed the Romans — needed the administrators, the churchmen, the local landed elite. When, after Constantine, Christianity effectively became the empire's official religion, the barbarians proved quick to adopt it. At mass in the Pantheon in Rome one evening last year, I was surprised to hear the service conducted in German, for a church group from Bavaria, and I couldn't help recalling (*mea culpa*) that the Visigoths were already Christian when they sacked Rome in 410 A.D.

If there was a tipping point, a factor that made the barbarians a fatally destabilizing force within the Western empire, it was not so much their sheer numbers as the manner in which some of them were eventually let in — a manner that made their absorption far less likely. The migration of the Huns from central Asia pushed other groups into the empire along a broad front on two important occasions, first in 376 and then in 406, and the Romans permitted some of these groups to stay and settle in designated regions, under their own leaders and with their own active armies. This was a major break with long-standing policy. What was the rationale for allowing these autonomous enclaves of *foederati*? The Romans seem to have been strong enough to fight them off. Popular impressions to the contrary, as Whittaker points out, except for Adrianople "no major defeat was suffered by the Roman army anywhere in the Roman world between 350 and 500." It could simply be, as another historian observes, that "concessions to barbarians were safer than the domestic risks of efficient defense." Safer, for instance, than demanding more money and men from unaffected areas that were already drained of both. Safer, perhaps, than handing victories to generals who might then harbor designs on the purple.

Whatever the impetus or the benefits, this policy eventually brought on a crisis. It meant giving up the revenues from wherever barbarians settled in Gaul, Spain, and Africa. And it brought into the empire significant new blocs that were neither one thing nor the other, neither outright barbarians nor integrated Romans but some sort of second-class citizen — some sort of limbo entity. They had armies, though, which made them players in the empire's already contentious politics, to a degree that gave imperial succession and imperial governance the character of a chaotic charade. Eventually the autonomous regions would devolve into *de facto* kingdoms, seeing no need to maintain the pretense of participation in some hollow imperial enterprise. A well-known summing up by Walter Goffart both exaggerates and makes a telling point: "What we call the Fall of the Western Roman Empire was an imaginative experiment that got a little out of hand."

Many Borders, and None

ONE LESSON OF ROME'S FRONTIERS is that borders aren't as simple as a line in the sand, or in the bog, even when you want them to be; they weren't then, and they surely aren't now. (Even the Rhine isn't what it used to be: environmentalists lament that the river has been invaded by barbarian hordes in the form of alien species of crabs and salmon.) Any sprawling state has a variety of borders—and not all of them lie in exactly the same place. There may be a political frontier for administrative purposes, but the economic and cultural frontiers can stretch far beyond it, as is the case with Britain; or, alternatively, they may not extend convincingly from the center to the outer borders at all, as is the case with Russia. The demographic frontier may wind in and out of political frontiers, as in the Balkans, or consist of an archipelago of population islands, as in Palestine. In diverse societies the religious frontiers may be impossible to map in any conventional way, but they constitute serious flashpoints nonetheless. Disease frontiers may exist in the form of millions of thin pieces of latex. When a political border runs between societies that are at vastly different stages of development and are vastly unequal in terms of wealth and power—as between Rome and *barbaricum*, or between the United States and Mexico—the attempt to enforce separation is an unnatural act: the economic dynamic pushes the relationship the other way, toward intimacy, no matter how loudly either partner cries "No!"

A study was conducted some years ago in which respondents were given blank outlines of the continental United States and asked to delineate the boundary of the region known as the Midwest. The responses were all over the map, with minimalists drawing a core crescent centered on the Rust Belt, and maximalists indicating an expanse stretching from the Appalachians to the Rockies. So where, exactly, on our mental maps are the various external boundaries of America?

In cultural terms, of course, these boundaries are virtually nonexistent. Rome's culture was transported up river valleys and along other trading routes, but Roman influence had its limits. America's cultural output, in terms of entertainment alone, is available in high volume to anybody with a satellite dish or an Internet connection, a radio or a movie ticket. Everyone has a story. In Ravenna not long ago I emerged from the Basilica of San Vitale, a place of subdued mosaic splendor built under the Ostrogoths, and on the street outside encountered a shriveled old woman draped entirely in black watching *Bob Newhart* reruns on a TV in her lap. A friend told me once about entering a mud hut in the middle of the Sahara Desert and seeing *The Cosby Show* flickering in the background. America, in turn, is infinitely porous, open to invasive influences. There are still a few countries (North Korea literally, France symbolically) that try to maintain an old-fashioned regime of cultural integrity, but in a globalized world, cultural borders are hard to police.

For the Romans, military and administrative borders were not always in precise alignment. There were forward bases in areas well beyond the territory Rome actually sought to administer. America has taken this to an extreme. Yes, America has a Coast Guard to watch its shores (and also to patrol the Mississippi and the Great Lakes), but for most of our history the sea itself was the most reliable bulwark. In recent years somewhat more attention has been given to protecting key U.S. ports against terrorism, much as the Romans fortified the "Saxon Shore" of Britain. The American defensive posture is focused on preventing attacks from happening in the first place, and on protecting far-flung economic and strategic interests as well as the territorial heartland. America's military deployments are not remotely defined by its geographic boundaries. The Distant Early Warning system—the network of tracking stations designed to warn of impending nuclear attack, better known as the DEW line—is strung across the top of Canada. So in this sense America's northern border runs not through Minnesota's Lake of the Woods but through the permafrost of the Arctic. At any given moment at least half of the nation's twelve carrier battle groups are out at sea — in the Pacific, the Mediterranean, the Indian Ocean, the Persian Gulf; America's security borders run across their flight decks. Defensive barriers now surround every American embassy in every world capital, obvious in their camouflage, like a bad toupee. And it's necessary to ask a question that never occurred to the Romans: How high in the heavens do our borders go? America's communications and data-collecting satellites, its first line of defense, fly at various altitudes, up to geosynchronous orbit at 26,200 miles. They're like Rome's *frumentarii,* in a way: they can gather information about crops *and* listen to conversations.

In terms of security, it's hard even to say where the border of a place like New York City actually lies. To be sure, there's the Hudson River and Arthur Kill on the west, forming a moat against New Jersey; and the bay on the south; and various administrative boundaries on the north and east; but the reach of the New York City police force now extends far outside the city's official boundaries, to offices in Russia and Israel, Britain and France, Canada and Singapore. NYPD officers today can be found almost anywhere, from the docks of Rotterdam to the bazaars of Istanbul to the warehouses of Dubai. More than a thousand officers work the counterterrorism beat, as compared with a few dozen before the attacks of 9/11. New York is deploying its local *limitanei,* its border auxiliaries, along a *limes* of its own definition; it's not depending on the imperial legions. As one account explains:

> There was a strong feeling that federal agencies had let down New York City, and that the city should no longer count on the Feds for its protection. Some of [Police Commissioner Ray] Kelly's initiatives were incursions into territory nor-

mally occupied by the F.B.I. and the C.I.A. And yet few objections were raised. It was as if the Feds, reeling from September 11th, silently acknowledged New York's right to take extraordinary defensive measures.

Reading those words, you may get some feel for the worldview of a Roman magnate in distant Gaul in the fourth or fifth century, who sees that Rome is increasingly far and threats are increasingly near, and that perhaps the time has come to build some relationships with other groups that wield power, and to fend for himself.

If you think of America not as a political or demographic entity, whose population shares certain values and viewpoints, but as mainly a collection of economic interests, then your idea of where America's boundaries lie needs to shift again — and to shift inward as well as outward. Actually, here's where those old-fashioned textbook maps of the barbarian invasions prove graphically suggestive, though the invaders need to be relabeled: they're not the Suevi and the Alani and the Franci but, rather, groups like Siemens, Royal Dutch Shell, British Petroleum, Deutsche Bank, Cadbury Schweppes, and Novartis. The past few decades have seen an unprecedented influx of foreign companies seeking to take over American ones. Chrysler was bought by Germany's Daimler-Benz, Random House by Germany's Bertelsmann, Mack Trucks by France's Renault, and Columbia Pictures by Japan's Sony. Ralston Purina, deep in the Great Plains, was overrun by the Helvetii — it's now owned by Nestlé, in Switzerland. From time to time a major takeover bid is stopped, as in 2005, when an attempt by the Chinese national oil company to buy the American oil-and-gas company Unocal ran into political opposition from nearly every quarter. The House of Representatives passed a resolution stating bluntly that Chinese ownership of an important American energy company would "threaten to impair the national security of the United States." In the end the

Unocal deal did not go through. This isolated episode, though, is perhaps equivalent to one of those Roman victories that failed to turn back the tide of history — the Battle of Chalons, say, where Aetius deflected Attila. By then the damage had already been done: the Huns had driven other tribes into the empire, and those other tribes would not be leaving.

If you're of a mind to entertain a dystopian fantasy parallel, it's not hard to envision how a modern Asian catalyst could produce an analogous, Hun-like ripple effect. Were the Chinese to begin unloading the enormous amounts of American debt they hold, one consequence would be a significant and rapid decline in the dollar. The inevitable aftershock of that would be more and more American companies falling into outside hands, and small-scale "barbarian kingdoms" sprouting up from Tacoma to Marietta, Long Beach to New London, with American workers by the millions answering to foreign managers, and local governments looking eagerly for ways to establish a *modus vivendi*. In textbooks a thousand years hence there might be gatefold maps of America "at its greatest extent," with little crossed swords showing where, say, the Chinese were deterred at the gates of Maytag, or the United Arab Emirates were held off at the mouth of New York Harbor. But mostly there would be sweeping arrows of corporate conquest, with little dates marking each successive line of advance. And the caption might mention how the English spoken in Alabama had been influenced by Korean, or how all the slang terms in Missouri for "efficiency" and "downsizing" were loanwords from Japanese.

That's the nativist nightmare, at any rate — or one of them. The reality is that "borderlessness," *de jure* or *de facto,* is a fact of life, and Americans are perpetrators of it as much as victims. We ourselves play the role of barbarians in other places. Borderlessness cuts across all categories — culture, economics, demographics. The biography of a counterfeit Prada handbag serves as a parable of a borderless world:

The bag's original design — probably acquired or stolen in Europe — was transported electronically or physically to China. There, the leather, zippers, belts, and buckles were procured and assembled into tens of thousands of counterfeit handbags. The finished products were then smuggled onto containers officially carrying, say, industrial valves, to ports such as Naples or New York. Once the handbags reached these final markets, street merchants took over — often African immigrants who themselves were smuggled across borders by human-trafficking networks.

In the final act, the profits from the enterprise are laundered electronically and deposited "offshore" in untraceable numbered accounts in the Cayman Islands.

The Assimilation Machine

THAT PRADA HANDBAG has no ready analogue in the Roman experience: the modern world is on its own. America's physical borders are another story; here the resonance with Rome is substantial. The U.S.-Mexico border is our Rhine and Danube frontier. It is about the same length as that segment of the Roman perimeter, running nearly 2,000 miles along the bottoms of four states and through the territory of three dozen native tribal groups. Where the border is not river, it is mainly desert or mountain. Once again the speakers of a Latinate language bump up against the speakers of a Germanic one, although linguistically the roles of *imperium* and *barbaricum* are reversed. Some 10,000 Border Patrol officers — two Roman legions — are charged with the task of keeping people from crossing *la linea*. But by any practical measure this border is as indefensible as the Roman one came to be — it is too long, and the terrain too inhospitable, and the guards too few. As was the case with Rome and its neighbors, the wealth on one side is also just too great; according to the historian David Kennedy, the wage differential between America and Mexico "is the largest between any two contiguous countries in the world." By now the broad zones on either side of the border, fifty or a hundred miles wide, resemble each other far more than they do the distant cores of their own countries. There are 300 to 400 million legal crossings of the border a year, many by people whose work takes them back and forth every day. And then there are those crossings, legal and illegal, made by people who plan on longer stays—that is, entire lives—in the north.

At El Paso the white SUVs of the Border Patrol stand sentinel at half-mile intervals along the banks of the Rio Grande, as if unaware that history has long since passed them by, heading in both directions. El Paso itself is already 90 percent Mexican. The Wells Fargo Bank and other tall downtown buildings stand like Anglo boulders above a Hispanic tide. Of course, it's not just from Mexico—the *oikumene,* the known world as it registers on the American consciousness, keeps getting bigger. The old dynamic of Rome is familiar to Americans. Political and economic dislocations far from our shores—at first in Europe and Latin America, more recently in Asia and Africa—have set people by the millions into motion, and in our direction. America's population would be in decline were it not for this replenishment from the outside. California, home to 40 percent of America's immigrants, where César Chávez's birthday has joined the Fourth of July as a legal holiday, now has the demographic character of a Roman frontier province. The classicist Victor Davis Hanson, in his book *Mexifornia,* sees a "potentially explosive mix" in this mingling of races, cultures, and classes—"a demographic and cultural revolution like no other in our times" (but one not unknown in the Rome of the Caesars). Is this blending getting a little out of hand? Immigrants are choosing new destinations, settling down in out-of-the-way places. A restaurant in my small New England town—a place still governed by town meeting, and where the barbers function as newspapers—offers something called "Thai *fajitas*" on its menu. In the past five years the number of immigrants in South Dakota has grown by 44 percent, in Delaware by 32 percent, in New Hampshire by 26 percent. Whittaker, the historian of the Roman frontiers, takes the long view and makes this analogy:

> The Roman empire of the fourth century was in some ways undergoing the same kind of transformation as the modern nation-state in the face of globalization. Both can be viewed as what Karl Barth called "disordered societies"; that is, as societies where traditional values were in conflict with new interests, when relations between national and foreign cultures were being renegotiated, and when the concept of ethnicity was being redefined under pressure from external frontiers.

As a national issue, immigration is politically pyrophoric—it has a way of igniting for no apparent reason. One day all eyes might be on a coming election or an endless war, and then suddenly the focus shifts to Minuteman vigilantes patrolling the Arizona and New Mexico border. Frustrated that the borders are so hard to control, and that the world is not a tidier place, American politicians have proposed tough measures to seal the frontier and to make it a felony to be illegally on the wrong side of it. One California congressman has suggested that America's incarcerated population be pressed into service to replace undocumented workers: "I say, Let the prisoners pick the fruits." (Maybe he had Louisiana's "rent a convict" practice in mind; it provides certain prisoners to local businesses as day laborers, for low pay or for none.) The release of a Spanish-language recording of "The Star-Spangled Banner" in 2006 made talk-radio tempers flare—"The Illegal Anthem," one commentator called it. A letter to a columnist in the *El Paso Times* expressed concern for, among other things, the fate of what he regarded as traditional American food: "The images that sprang into my mind with the idea of the National Anthem in Spanish was a man coming into my home, sitting in my chair, changing my TV to Telemundo, my radio to KBNA, and announcing to my wife that the lasagna we were having will now be changed to menudo and burritos." Many in Congress are violently opposed to enabling the 10 million or so illegal aliens in the country to take steps toward citizenship—leaving them in permanent second-class status, a deliberate limbo, needed but not wanted: economic *foederati*. President Bush has announced his intention to erect real fencing and also a high-tech "virtual wall"—with infrared and motion sensors, and unmanned blimps and drones—along hundreds of miles of the U.S.-Mexico border. There is certainly money to be made in border defense. The Department of Homeland Security has recently solicited bids from vendors capable of supplying an "indefinite quantity" of bean-and-beef burritos for illegal aliens awaiting repatriation. (A hundred years ago, of course, the order might have been for lasagna.) The real money is in the virtual wall itself, to be built and operated by private corporate security forces that are supplementing the federal Border Patrol as America's *limitanei*. The irony is irresistible: using overpaid hired corporations to keep out underpaid hired workers.

In the long history of efforts at immigration reform, the biggest consequences are usually the unintended ones. What never changes, because in the short term it can't, is the underlying dynamic: a rich country's demand for unskilled workers, and the rest of the world's limitless supply of them. Like the Romans, Americans seem to know in their hearts, for all the hot rhetoric and partisan posturing, that there's little to be done: the forces of population and economics are too powerful. In the late fourth century the Romans carved public images of barbarians being slaughtered—and at the same time produced the medallion of Lyon, showing barbarians being welcomed as settlers. The President Bush who wants to fortify the border against the immigrant hordes also produced what might be called the medallion of San Antonio: a campaign film in which he puts one arm around a brown-skinned immigrant child and waves a small Mexican flag with the other. Maybe the problem isn't as much of a problem as would be any "solution" that restricted the supply of cheap labor. This is the modern version of "concessions . . . were safer than the domestic risks of efficient defense." Or, as the former senator Alan K. Simpson once put it, "We've done open arms and closed arms. Now it's 'throw up your arms.'"

But there is positive news for America in the Roman experience. It's impossible to do a precise accounting, but the outside pressures on Rome were at least as great as those on America, and in all likelihood far greater; and Rome coped with them successfully for a very long time — for centuries on end. There was hardly a moment in Rome's history when barbarians weren't worrisome neighbors. Rome's ability to assimilate newcomers is so large a fact that it's easy to lose sight of. Attention is drawn away, understandably, by specific episodes of invasion and mayhem in the final centuries of the Western empire's coherent life. "They gave vent to their rage by every kind of atrocity and cruelty," runs one account of the Vandals in Africa. And so they did. But the peaceful influx and evolution over great swaths of time was a bigger phenomenon still: "In the many centuries since it had entered upon the conquest of distant lands," Walter Goffart observes of Rome, "millions of barbarians had been pacified and absorbed into a common civilization, a Romanita whose component peoples, however imperfectly homogeneous, looked to the emperor for defense against outsiders, and had no desire for liberation from his rule." These many millions of barbarians were incorporated into Roman society as a matter of routine, and without the help of powerful tools Americans take for granted, such as public education and mass communications (and Madison Avenue). There was no conscious effort to encourage assimilation as such, no set of policy initiatives, no Department of Romanization. But it happened, on a broad scale, because in the eyes of the newcomers Roman civilization looked like a good deal and they wanted to buy into it. Think of those many hundreds of scribblings pulled from the muck of Vindolanda. This is the detritus of everyday life, produced mainly by soldiers who by birth were not Romans but barbarians, and rather unsophisticated, and still more comfortable speaking in their native tongues. They were willing to invest twenty-five years in the military in order to obtain Roman citizenship for themselves and their families; half of them would die before achieving that goal. The Latin they wrote is clumsy, filled with misspellings and odd locutions and unusual turns of mind. But it is Latin, real Latin.

America, too, is an assimilation machine, though one whose success we tend to acknowledge mainly in hindsight. Looking back, we now know that America managed to accommodate the waves of immigration in the 1850s, the 1880s, the first decade of the 1900s, and the 1980s, despite skepticism at each of those moments — when those moments were "the present" — that it ever could. In the earliest days of Ellis Island, before the stricter immigration laws of the 1920s, America's door was open wider than it is today; some 98 percent of those getting off the boat were allowed in, virtually without paperwork; and most were processed within eight hours of arrival. Every age doubts that it retains the absorptive capacity of ages past, just as every age fails to remember the human heartache and wrenching adjustments that immigration entails. Or the utter determination. My father-in-law came to America from Mexico in 1920, in his mother's arms, and on his yellowing immigration papers, next to "Mode of arrival," is the typed-in word "Rowboat." My children, now that I think about it, have the kind of mixed heritage that would have been commonplace on Rome's frontiers: one Mexican ancestor was coming north to become American just as one Irish-American ancestor was heading south to fight Mexicans.

The forces at play here are powerful. Politicians can tweak the laws, but "policy" is a microclimate compared with the swirling vortex of the American economy and American culture. In Los Angeles and elsewhere, Spanish-language radio stations are going bilingual or shifting from Spanish to English, because that's what the younger audience wants; 90 percent of Hispanic high school graduates prefer English to Spanish. There is a reason why Islamist cells are sprouting all over Europe among the Muslim underclass but have gained little traction in America. An expert on the subject, Robert Leiken, says, "If you ask a second-generation American Muslim, he will say, 'I'm an American, and a Muslim.' A second-generation Turk in Germany is still a Turk, and a French Moroccan doesn't know what he is." For second- and third-generation Asians and Hispanics, the rates of intermarriage with other groups in America now range above 30 percent. Population projections suggest that by 2050 one American in five will be of mixed racial ancestry. A professor at California State University says, "We don't know how it will affect who our Miss America is twenty years from now, but it will." Another analyst, who studies the "great immigrant portals" of Houston, New York, and Los Angeles, writes, "Among the people working there, concepts such as 'ethnic solidarity,' 'people of color,' or 'cultural community' generally count for less than basic principles such as 'Does this sell?,' 'What's my market?,' and ultimately 'How do I fit in?'" Reading those fragments from Vindolanda, I'm reminded of the printed cards passed out on the Mall in Washington in 2006, where thousands of Hispanic immigrants united to say these words:

> *Ai pledch aliyens to di fleg*
> *Of di Yunaited Esteits of America*
> *An tu di republic for wich it estands*
> *Uan naishion, ander Gad*
> *Indivisibol*
> *Wit liberti and yostis*
> *For oll.*

When you step back and pose the civilizational question about Rome — On balance, who prevailed? — it's not at all clear that the Romans were driven from the field. The religion of the

Romans, Christianity, became the religion of the newcomers, and to this day the language of the Romans is, in a sense — *mutatis mutandis* — spoken by their descendants. The people in the Roman lands remained drinkers of wine. Their architectural designs and building techniques became standard. Their towns and cities are still inhabited, and their seasonal rituals, under different names, still celebrated. Their attention to law and to legal systems spoke powerfully to America's own Founders, and Roman law remains foundational in Europe to this day. The notion embodied in imperial Rome — creating unity out of diversity — has been a grail of geopolitics ever since. The Roman Empire disappeared, of course, as a formal construct, but in other respects it did not entirely vanish. Indeed, how could it simply vanish, a thing so large and pervasive? Where would you put it? In a masterly and influential study published several decades ago the historian Peter Brown took up this very subject. Brown was born and educated in Ireland, and witnessed firsthand the withdrawal of imperial order from his native land — its withdrawal, yes, but also its retention and persistence, its stubborn influence, the deep taproot of its legacy. The experience left its mark on him as a historian of Rome. Yeats had written romantically of rebellion as having left Ireland "changed, changed utterly," but Brown never warmed to the "changed utterly" school of historical thought. Years later, he conjured an expanse of time running from Marcus Aurelius, in the second century A.D., to Charlemagne, in the eighth, that he would call "late antiquity," a period of both continuity and transition, and of great cultural vitality. He was able to describe this period, as he later reflected, "without invoking an intervening catastrophe and without pausing, for a moment, to pay lip service to the widespread notion of decay."

His deliberately provocative question, in the end, is really this: Did Rome ever fall?

Epilogue
There Once Was a Great City

One thought alone preoccupied the submerged mind of Empire: how not to end, how not to die, how to prolong its era.
—J. M. Coetzee, *Waiting for the Barbarians*

T HE CITY OF RAVENNA grew out of the Adriatic marshes, as Venice one day would, rising on tufts of land and thousands of pilings. The city eventually became a major port, and a second base for Rome's fleet. Julius Caesar gathered his forces at Ravenna as he prepared to cross the Rubicon. Because Ravenna —surrounded by wetlands—was so easy to defend, the capital of the Western empire was moved there from Rome by the emperor Honorius, early in the fifth century. The mosaics of Ravenna, created around this time, still draw people to the city. So does the resting place of Dante, who died near Ravenna while in exile from Florence. His body was never returned to his native city. The elegant tomb in Florence's Church of Santa Croce is empty.

Ravenna was also the scene of the Roman Empire's final act —or, rather, its final act in the West. The figure regarded as the last in the long imperial line, Romulus Augustus, spent the few months of his reign in Ravenna, in 475–476 A.D. He was a boy of no more than thirteen or fourteen. Because he was so young, people gave the name Augustus a diminutive twist — "Augustulus," the emperor was called: "Little Augustus." And because he was both a puppet and a usurper, people also gave the name Romulus a pejorative cast — "Momyllus," they said, or "Little Disgrace." There was time enough in his reign to mint money, barely, although the young emperor's name was so long that it was hard to fit onto coins. (The emperor of the East at the time, Zeno, did not have this problem.) A number of buildings still stand in Ravenna that date back to the time of Romulus Augustulus, or even earlier. The young ruler would certainly have known the octagonal baptistry, with its extraordinary ceiling mosaic depicting John the Baptist pouring water over the head of a naked and beardless Jesus. Looking up at it, you may wonder if the beardless boy-emperor, himself only recently elevated, saw in the image anything ironic or apt about his own circumstances. The sources do not indicate whether he was self-aware.

The sources, in fact, indicate very little. Rome left behind a great deal of itself, in stone and metal and writing — more than any other ancient civilization — but it's useful to remember how modest "a great deal" can be. In the library where I've been working, the Boston Athenaeum, you can find nearly every written word that survives from ancient Rome — history, literature, philosophy, inscriptions — and nearly every word published in English about Rome by anyone in the past three centuries. It's mostly shelved in an adjunct vault of the library, five stories high, called "the Drum." The aisles are narrow and the ceilings low, and there are no windows; it calls to mind a Roman *columbarium*, where cinerary urns were packed tightly in niches. My own calculation, extrapolating from the Drum, is that you could pack all the writing from a thousand years of Roman history, plus all the writing about those thousand years, into a cube about forty feet on a side — roughly the amount of space taken up by White House documents from just the three years of the Kennedy presidency. Ramsay MacMullen, with typically elegant understatement, has referred to "the extreme recalcitrance of the evidence." A curiosity seeker looking to answer mundane questions about Rome can drill a lot of dry holes. The twenty or so "barracks emperors" each ruled the world for a few months or years in the mid third century, but often historians have to guess at their precise dates. We know that Nero, famous for playing the cithara while Rome burned, in fact cared deeply about his music and entered public competitions; he would lie on his back with weights on his chest to increase his lung capacity. But we don't know basic details about the imperial government and budget during his reign. The surviving record about Rome is not only reticent but capricious. Imagine if America were to leave behind for historians the complete text of the Starr Report but only the 1947 volume of *The Statistical Abstract of the United States*.

What we know about Romulus Augustulus is this: He was the son of a senior officer named Flavius Orestes, an aristocrat from the provinces, half Roman and half German. When Attila's power was at its peak, Orestes served the Hun as a secretary and a diplomat. It was a time when allegiances could shift rapidly. After the death of Attila, Orestes was appointed *magister militum,* or master of the Roman armies, by the Western emperor, Julius Nepos. The imperial armies in Italy by now consisted mostly of barbarians, and Orestes, with his German background, enjoyed a presumption of favor. Once in control of the military, Orestes deposed Julius Nepos and installed himself as the *de facto* head of government in Ravenna. He declared his son rather than himself to be the new emperor, perhaps because Romulus, with a fully Roman mother, might seem to have more of a claim. All went well until the German troops demanded that Orestes make good on a promise to give them land. He refused, and faced a revolt led by a senior officer named Odoacer. Orestes took refuge in what is now Pavia, but the city was captured, and he was taken away and beheaded. Odoacer then marched on Ravenna and deposed Romulus. The date was September 4, 476 A.D.

What happened to Romulus? His life was spared, surprisingly. We know that he was exiled to the western coast of Italy — to Neapolis, today's Naples. He seems to have left Ravenna with his mother, and no doubt with a large retinue, and he was even given a pension of 6,000 *solidi* a year — a great deal of money (according to one estimate, the equivalent of forty kilograms of gold). The most obvious route into exile would have been the Via Flaminia — roughly the path of the current Strada Statale 3 — to Rome, and then the ancient Via Appia, already 700 years old, south to the vicinity of Naples.

That whole stretch of the Italian shore was vacationland for the Romans. Museum drawers are filled with ancient beach-town baubles of glass or clay: "Souvenir of Neapolis," they might as well say, or "This Mule Climbed Mount Vesuvius." Villas crowded the lush volcanic hillsides. Sluice gates brought the renewing sea into the teeming fishponds that each great estate would have; the truly rich were known as *piscinarii*, "the fishpond set." In a seaside town called Baiae you can still pick your way, more or less alone, through the ruins of the emperor's summer home, with its mosaic floors and its long terraces on four or five levels. Hadrian died somewhere on this hillside. Plutarch has described the estate of Lucullus, the general and statesman of the late republic, on the Bay of Naples — an estate supporting a way of life so opulent as to give rise to the term "Lucullan." Gardens cascaded luxuriantly down to the sea. Offshore follies floated on the waves. The food and drink were renowned for their quality and plenitude, and were served not merely to impress others. Once, when Lucullus was dining alone, a servant prepared a simple meal for him, and was told by his master to go back and bring forth something more extravagant, as if there were guests: "Did you not know that tonight Lucullus is dining with Lucullus?"

In some diminished condition the estate of Lucullus must have survived into the fifth century, because Romulus Augustulus apparently passed his days on part of it, on a headland jutting into the bay. A monastery would grow on the site, enshrining the bones of Saint Severinus; strong circumstantial evidence suggests that it was founded by Romulus and his mother. If this is the same "Romulus" to whom the Ostrogothic king Theodoric wrote a letter in 510, then Romulus Augustulus lived for a good long while after being run out of Ravenna. Odoacer was not so fortunate. His armies were defeated by that same Theodoric, in 493. Theodoric then invited Odoacer to a banquet—no hard feelings—and killed him with his own sword.

Medieval fortifications now crown the Lucullan headland; the complex is called Castel dell'Ovo, the Castle of the Egg. Legend has it that the poet Virgil placed an egg in a secret room underneath the castle, and that if the egg is ever crushed, Naples will be destroyed. Looking at Naples today, it's hard to deduce whether the egg has remained intact. The headland divides the tolerable tourist city to the north from the grittier seaport city to the south. But the breeze off the bay is fresh and constant. Even without vendors selling *granita al limone* it would have been a congenial spot in which to endure your exile, especially on 6,000 *solidi* a year and with Vesuvius quiet. For many Roman emperors, the end had been far less kind, and the breeze far more fleeting, and felt only on the back of the neck.

Passing the Scepter

WHETHER THE ROMAN EMPIRE came to an end in 476 or some other time hardly matters—at some point it came to an end. Is "fall" the right word to describe what happened to it? And is understanding what happened to it the right way to think about what may happen to America? When the communist system in Eastern Europe expired, in 1989, and the Soviet Union began to splinter apart, many observers warned of the imminent collapse of Soviet society. Others made the sensible point, since confirmed by events, that the Soviet organism was too massive, sprawling, and dense to undergo something as sudden or straightforward as "collapse." It was the world's biggest producer of oil and gas, of iron ore and steel. Its vast and growing underground economy showed that entrepreneurship was very much alive. Soviet Russia was, as one observer memorably put it at the time, "too big to fail" in any conventional sense, even if the chaos of unpredictable mutation might be its lot for decades.

The Roman Empire was also too big to fail, if by "fail" you mean "suddenly discontinue and disappear." Its methods of agriculture, its patterns of trade, its cities and ports, its buildings and infrastructure, its modes of administration, its names for objects and places, its laws, its elites — to varying degrees in various places all of these things lived on, for longer or shorter periods, making the path forward an irregular transition rather than a catastrophic revolution. The historian Lynn White states simply, "There is, in fact, no proof that any important skills of the Greco-Roman world were lost during the Dark Ages even in the unenlightened West." Rome dissolved unevenly into history, vanishing dramatically in only a few areas (first in Britain, from which imperial forces unilaterally withdrew at the beginning of the fifth century; and then, 200 years later, in the regions that came under the sway of Islam). In many ways, for good or ill, a version of Rome was carried forward into new eras and places by the Catholic Church — its language, many of its values, much of its administrative structure, some of its dress. You can still see Rome clearly in the Church's organizational charts, from "pontifex" through "diocese." Maybe someday the Mormon Church, the fastest-growing religion worldwide, will play a similar role, encapsulating and propagating a particular, canonical version of America — its middle-class striving, its family values, its missionary outlook. It's not a far-fetched notion: Salt Lake City as the Vatican of the third millennium.

The unraveling of Rome was undeniably a big political event:

a great unity was irrevocably diminished; a great and wondrous order became a thing of the past. The Romans saw it happening slowly before their eyes, wrote about it, and openly debated the reasons for it — you will find no "declinists" more sour than Latin writers of the fourth and fifth centuries. They did not agree on the causes any more than modern specialists do. But ordinary life did not change suddenly in most places. One historian writes, "After the initial shock of barbarian incursions in the fifth century, the separate localities of the Roman Empire passed rather easily into barbarian hands. Landlords continued to manage their properties; peasants worked the land; and members of the imperial bureaucracy fulfilled their functions — only now in the service of barbarian tribes and chieftains rather than of Roman emperors." A senate composed of aristocrats will continue to exist in Rome for more than a century after Rome's fall. Yes, the archaeological record reveals a general subsidence in well-being over time. Pottery is less well made. Houses are more likely to be built of wood than of stone. Fewer villas boast mosaics and plumbing. The money economy contracts. Livestock lose weight. But for generation upon generation, in the aftermath of the empire as before it, life was for most people what it always is: a series of incremental adaptations that only the passage of centuries reveals to have been a radical departure, or to have been pointing in some clear new direction.

The Romans had seen themselves as the successors of the Trojans and the Greeks — indeed, as the culminating empire for all of time: "empire without end." Of course, the example of Troy was also a sobering one. Virgil was speaking of Troy, not Rome, when he recalled the demise of a great city — *"urbs antiqua ruit"* — but the words cast a shadow of foreboding all the same. When the Roman Empire was truly gone, European rulers invoked Rome's mystic authority for themselves — the "translation of empire," as the process came to be called. Slavic peoples saw the imperium passing from Rome to Constantinople and thence to Moscow, which they called the "Third Rome." Western Europeans traced the *translatio imperii* through the Holy Roman Empire and up to their own doorsteps. English imperialists at the height of the British Empire were full of talk about inheriting the Roman mantle, and when the mantle began to slip, they knew full well who would pick it up. Bishop Berkeley, writing in the mid nineteenth century, looked to America: "The world's scepter passed from Persia to Greece, from Greece to Italy, from Italy to Great Britain, and from Great Britain the scepter is today departing. It is passing to 'Greater Britain,' to our mighty West, there to remain, for there is no further West." (The city of Berkeley, in what was thought would be the "no further West" state of California, is named for this same bishop.) Charles Darwin joined in, giving the idea biological sanction. Endorsing a book that painted America as the culmination of history, Darwin wrote, "There is apparently much truth in the belief that the wonderful progress of the United States, as well as the character of the people, are the results of natural selection." With the transfer of the imperium to America, the British found solace in comparing themselves to Rome's more cultivated Greek antecedents, who never ceased looking down on the upstart that overwhelmed their civilization.

There has always been a whiff of determinism about the Rome-and-America analogy, as if the fate of Rome foretold the fate of all future empires. Byron's Childe Harold, contemplating the city's ruins, captured a famous version of this outlook:

> There is the moral of all human tales;
> 'Tis but the same rehearsal of the past,
> First Freedom, and then Glory—when that fails,
> Wealth, vice, corruption—barbarism at last.

But press it too far, or invoke it too literally, and the Rome-and-America analogy breaks down in strategic places. Rome accepted and bestrode its destiny. Americans don't yet agree that an empire is what we've become, much less agree that we ought to be one. The political gulf between Rome and America is wide, morally and procedurally. America's democratic form of government looks to us like a flawed and tarnished thing, and we lament its grave deficiencies. But it's more adaptable, just, and robust than anything Rome came up with in a thousand years. Elections remain a check on power, a crude and clumsy but as yet sacred way to reorient the compass.

Our judicial apparatus at every level depends increasingly on the ability to pay large sums for lawyers—implicitly creating a two-tier system of justice, which gives us pause. But this system does not represent an official ideal; many Americans fight against it. Rome was explicit and unabashed about its official two-tier system: there was a small group of people called *honestiores,* the privileged elite, who enjoyed a presumption of integrity and were exempt from many judicial penalties, and from interrogation by torture; and then there was everyone else, the *humiliores.*

Americans are well aware of the nation's worsening income inequality. The wages of ordinary working people have in recent years become detached from rising productivity and allowed to sink, in real terms, even as America's *piscinarii,* the small group at the very top, see their share of the nation's total wealth grow and grow, and their taxes shrink and shrink. But most Americans don't want our society to be like this, and we remain at heart a middle-class nation; whereas Rome lacked what we would regard as a middle class, and the very tiny Roman elite accepted the chasm between themselves and everyone else as the divinely ordained natural order and an affirmation of their own virtue. Pliny the Younger observed, "Nothing is more unfair than equality."

The public savagery of Rome would be a shock to Americans. Leave aside the gladiators and the carnage of the arena. Leave aside the practice of hanging the heads of executed public figures, including Cicero, from the Rostra in the Forum. In city and country the rich got their way in the manner of warlords and mafia dons, using retainers and "men of the arm" to beat and torture and kill; Ramsay MacMullen has written about the "predatory arrogance long latent in the pax Romana" and "an uncontested right to assert one's superiority in offensive, often cruel, ways." Americans lament but tolerate the impersonal, structural savagery of economic forces; they regret but accept the distant cruelty of war's "collateral damage." But Americans cannot imagine the routine interpersonal oppression that Romans knew and took for granted.

And then there is the enormity of slavery, which America eventually rejected but Rome never did. The distortions and habits of mind introduced by slavery make Rome an alien place. For centuries Rome ran on slaves the way America has run on oil, and wars often brought a significant increase in Rome's energy supply. At a stroke the conquest of Epirus, across the Adriatic, in 167 B.C., brought 150,000 slaves to Rome; Caesar's wars in Gaul, a century later, brought a million more. On the great estates, slaves vastly outnumbered free men; in Italy at the end of the republic, slaves may have made up half the total population. The revolt led by Spartacus showed what a terrifying threat these slaves represented, and Roman society was always mindful of the need to keep the bondsmen in bondage. The statutes were harsh. One law stipulated that if any slave killed his master, *all* the slaves in that household must be executed—this to discourage conspiracies and to give every slave a stake in the safety of his owner. In other words, the system as a whole was premised on mutual terror.

Could there ever be extenuating circumstances? Was Roman society secure enough to stay the law's hand? Both Pliny and Tacitus tell the story of a prominent politician named L. Pedanius Secundus, who in 61 A.D. was killed by one of his slaves. It appeared to be an isolated incident—a private grievance, not a conspiracy — and the idea of exacting the full measure of punishment gave many Romans pause. Pedanius had been a very rich man; this one household was home to hundreds of slaves. But in the end the Senate decided to apply the letter of the law and to execute them all. Only the certainty of punishment could keep the slaves in check, it was argued. One senator explained: "It is true that, because of this, innocent people will be sacrificed; but every exemplary punishment always contains an element of injustice, which is carried out on individuals in the name of the utility of the entire people."

America is far removed from the kind of society described here. Few Americans would have wanted to live in it, even if guaranteed a place at the pinnacle, with the wealth and status of a Symmachus or a Lucullus. Why, then, do we feel a tug of loss when contemplating Rome's demise? The answer is not to be found in Gibbon, who regarded "the tranquil and prosperous state of the empire" in its golden age as the apogee of human happiness. Gibbon was a man of his times, blind to many things, and in any case this sort of sweeping historical assessment is out of fashion. I remember reading an interview some years ago with the editor of a series of books about the British Empire, who was asked whether he thought that the empire, on balance, had been a force for good or ill. He sputtered at the question, the impossibility of the calculation — you could almost feel a fine spray of expostulated sherry. No, the tug of loss doesn't have to do with any grand cost-benefit analysis, or even a fleeting thought that we'd all be "better off" if the status quo had just gone on and on. What draws us now is something far more elemental and emotional: the brutal reminder of impermanence. That, and from time to time an anxious flicker of recognition — the eagle in the mirror — when catching sight of the characteristics that Rome and America share.

What's an empire to do?

Fast Forward

FOR REASONS THAT REMAIN OBSCURE, the poet Ovid was sent into exile by the emperor Augustus in 8 A.D. Ovid blamed his fate, cryptically, on "a poem and a mistake," and would say nothing more. He lived the last decade of his life on the shores of the Black Sea, pining for Rome. A long poetic work he wrote at this time is called *Tristia — Sadness*. It is intensely personal, and was probably intended to persuade Augustus to lift the sentence of exile. (He didn't.) In *Tristia*, Ovid imagines the journey his poem will take on its way to Rome: "Little book, I don't begrudge it; you'll go the City without me, / Ay, to the place where your master isn't permitted to go." As described in *Tristia*, the book arrives in Rome, enters the Forum, travels down the Via Sacra and past the Temple of Vesta, makes a right, passes through the gate of the Palatine, goes up the hill, and stops before its final destination, the house of Augustus. Over the doorway hangs the famous oak-leaf crown, awarded to Augustus in perpetuity by the people of Rome.

I had *Tristia* on my mind as I made my own way not long ago down the Via Sacra, past the Temple of Vesta, and through the gate of the Palatine. I paid an entrance fee in euros — the first official common currency in Europe since the days of the Roman Empire — and went up the hill, spending an afternoon under the umbrella pines and among the ruins. The house of Augustus is still there, modest in size, surprisingly well preserved; on the second floor is the elegantly frescoed room he referred to as "Syracuse," or his "workshop," his cherished private space, its window overlooking the Temple of Apollo. The Domus Flavia, the palace complex built by those who came after him, sprawls nearby, in some places several levels deep, in others a masonry footprint open to the sky. An apse in one chamber was meant for the emperor's throne; you can stand on the very spot — chew gum, use your cell phone — in this place that you would have ap-

proached with humility and awe, maybe terror, in 100 A.D. Behind the apse lie the remains of a cloistered garden area, the
peristyle. The mad emperor Domitian liked to walk here, and
had its walls veneered with high-gloss marble so that he could
see the reflection of any attacker. (It worked: he was stabbed to
death someplace else.) Just beyond the peristyle is the state dining room, flanked by fountains, with a double floor that allowed
heated air to circulate. The dining room leads to the family quarters, a private enclave set apart from the public areas.

The mental transference from Palatine to Washington comes
all too easily. Here in the palace, the Praetorian Guard would
have worn civilian clothes . . . as our Secret Service does. Looking south from the windows of his residence, an emperor would
have seen what the American president, looking south, also sees:
an obelisk, Rome's in the Circus Maximus, Washington's beyond the Ellipse. The president's desk, set in the curve not of an
apse but of an oval, is backed by his own lush peristyle, the Rose
Garden. Wander the hallways of the Domus Flavia or the West
Wing, and you come across libraries and briefing rooms and ath-
letic facilities, and hideaways where the leader can be alone. The
walls in both places display images of national mythology and
civic religion. Marble busts of previous occupants observe from
pedestals.

It will be a while, I hope, before tourists stroll among weeds
poking up through the Map Room and the Oval Office, or pose
before the scenic remaining columns of the South Portico. In
Rome today you see leathery men in cheesy centurion's garb
posing with tourists in front of the ruins. I'm not sure I want to
know what the Washington equivalent will be — Green Berets,
maybe, or TV reporters, or special prosecutors.

But the thought of tourists posing brings back the question:
Are we Rome? In a thousand specific ways, the answer is obviously no. In a handful of important ways, the answer is certainly
yes. As societies, America and Rome are built on different prem-

ises. As people, Americans and Romans cherish different values. But Rome and America share certain dangerous traits — habits of mind and behavior. America and Rome also face similarly fraught circumstances, arising both from inside and from outside.

From the vantage point of the far future — that is, projecting ourselves ahead — I imagine it will be possible to visit America in its olden days using computerized "fly-through" programs like the ones that exist now for ancient Rome, though far more elaborate and detailed. You'll be able to run the program forward and back — maybe watching the waters of the Potomac lap the South Lawn as sea levels rise, and then with a toggle watching the waters recede as time runs in the other direction. You'll be able to observe economic and demographic variables play themselves out over decades and centuries — the average American eye turning brown and almond-shaped; North Dakota becoming the first state with two senators but no electorate; the advent of prenatal sex selection teaching the virtues of randomness, but too late. Going the other direction, you'll be able to reverse-engineer "present-day" conditions, sorting out the tangle of social antecedents to determine what really caused what.

If you actually had such a program now, and could put early-twenty-first-century America into fast-forward, what would you see? Certain futures are all too plausible; we've made a start on each one of them.

For instance, there's the Fortress America scenario. As perceived threats to the country grow more insistent and varied, all of society increasingly bends toward a particular vision of homeland defense. We watch as local police forces, the educational system, even pop culture, bit by bit acquire a vaguely martial cast. Spending on domestic programs is diverted to national security. Economic life orients itself increasingly around the requirements of the military and the intelligence apparatus, and of our far-flung protectorates. Individual rights and freedoms take a back seat to the government's need to know. Borders are hardened. Privacy becomes just a footnote in the history books (though not in the ones used in Texas). The executive branch is paramount, the other two branches having evolved into useless but still-detectable appendages, like a whale's vestigial limbs. This would be Diocletian's empire taken to some future American extreme.

Or there's the City-State scenario, already emerging in many parts of the world. As the government in Washington becomes more and more unwieldy (and resented), and as its foreign policies drag the country into dangers that many of the country's components would just as soon avoid, the great cities gradually assert themselves. Los Angeles, New York, Miami, Seattle, Chicago—these and a few other places, including Washington, are America's prime source of wealth and creativity even now. They animate entire regions with their economic and intellectual power, and stamp those regions with their cultural character. Their leaders already do business with heads of state. They pursue domestic policies—on the environment, medical research, social issues—sharply at variance with those of Washington. As Richard Florida has pointed out, "If U.S. metropolitan areas were countries, they'd make up forty-seven of the biggest 100 economies in the world." Now, on fast-forward, we see them becoming *de facto* city-states, emerging organically out of the nation's moldering timber like the barbarian kingdoms of late antiquity. Of course, without the flywheel of Washington there is strife—over borders, resources, amenities. City-states compete even more ferociously than they used to for investment from abroad. Once, in a playful exercise, a friend took a map of the Northeast and drew what he thought should be the boundaries of the city-state centered on Manhattan, looping north to ensnare the Boston Symphony's summer outpost at Tanglewood. Would that be a *casus belli*?

And then there's what might be called the Boardroom scenario—the extension of corporate ownership to ever larger areas of ordinary life, not just in America but worldwide. The nation-state model of governance is only a few hundred years old. Why assume that it must be the last stage of political development? Why assume that America's government won't cede some or most of its social obligations to companies promising to "do it for us"? On fast-forward we see the rise of corporate feudalism on a global scale. The world's biggest corporations are already powerful transnational actors in an era when many problems demand transnational management. And what has just been said of cities can also be said of corporations: in this case, of the world's one hundred largest "economies," half are not countries but private companies. Some of them command small armies, and quietly rule significant swaths of the planet. Others manufacture the weapons used by "real" countries, including America. A small number of companies produce all the oil and gas. A small number control the world's freshwater resources. By means of privatization, big business is already acquiring — or being given, or being paid to take — much of the infrastructure of government. Social services may well be next. The capacity of corporations to do global harm is well established. They also have the capacity to do global good. Regardless, on fast-forward we watch them grow in relative power, untethered to any one sliver of national geography, but indisputable lords of the world's water, its food, its information, its health, its energy, its transportation, its software, its music, its security, its violence.

You can posit many possible futures, some deeply problematic, some relatively benign. Whatever comes to pass, the sheer fact of America will weigh on the world for millennia. Like Rome, America is in some ways inextinguishable. What we can't know is which characteristics will be extinguished and which won't. But we do have a say in the outcome.

The Titus Livius Plan

IF IT WERE SOMEHOW LEFT to me to figure out what ought to be nurtured or extinguished (this is what happens when you

stand in the emperor's apse too long), I'd start by trying to understand, and to acknowledge, what we're up against as a quasi-
empire in a turbulent century.

There is, to begin with, a psychological tendency that is
nearly impossible to shed. The idea that you should preserve everything you already have, exactly the way it is, exerts a powerful grip, even when logic suggests that only adaptation can preserve what is essential or worthwhile. "Herein lies one of the
curses of empire," writes the political scientist Eliot Cohen. "To
let go never looks safe, and indeed rarely is." It isn't only Americans who wring their hands after "losing" what they consider to
be strategic assets (fill in the blank: China, Vietnam, Iran, Iraq).
Britain's imperial devolution brought great bitterness. The emperor Nero was castigated because he "lost Armenia." Even wise
adjustments, at home and abroad, may be resisted.

Then there is a simple fact of life: the status quo never stays
that way. Thucydides observed that empires start to decline
when they cease to expand. You can't read an account of Rome
in the third, fourth, or fifth century, when expansion is over and
emperors are trying desperately to hold things together, without
marveling at the blizzard of variables in play. Every Roman action to address one urgent problem—military, diplomatic, economic, political—creates unintended new problems. The blood
thinner causes hemorrhage; the coagulant causes stroke.

And from this comes, finally, an unhappy generalization: large
systems are inherently unstable. There has been a lot of academic
theorizing in recent years about the nature and course of empires. An insight that transcends most others has been expressed
by many but was distilled into a few sentences by the economist
Paul Ormerod in his book *Why Most Things Fail:* "Species, people,
firms, governments are all complex entities that must survive in
dynamic environments which evolve over time. Their ability to
understand such environments is inherently limited. . . . These
limits can no more be overcome by smarter analysis than we are

able to break binding physical constraints, such as the speed of light." Unfortunately, there is no pause button.

For a long time Rome was able to organize the world according to its own convenience — until there came a point when doing so became difficult, and then impossible. The natural instinct is always to cling. In the late empire you see rich men like Symmachus trying to maintain the old habits and manners of Rome as if nothing had changed in two hundred years, or ever would. His modern editor comments, "We may say that Symmachus and his friends made a brave attempt, or we may say, in light of events, that they were blind." Americans, reading that assessment, may summon to mind a map with a little red arrow and the legend "You are here." Some would argue that it's well within our power, and perhaps also our duty, to keep juggling all the key variables. In the realm of foreign policy alone this could mean trying to prevent (as one analyst has written) "the rise of militant anti-American Muslim fundamentalism in North Africa and the Middle East, a rearmed Germany in a chaotic Europe, a revitalized Russia, [and] a rearmed Japan in a scramble for power with China in a volatile East Asia." That was a tall order even when those words were written — in 1998, before 9/11 and the wars in Afghanistan and Iraq added baffling new dimensions.

But what if it's not possible to control all the variables? What if Reinhold Niebuhr was right: the more powerful America becomes, the less control it exerts over its own destiny? In that case it would make sense to focus on the handful of big factors that are substantially within our control — and that contribute to social strength no matter what outside challenges we confront. The experience of Rome gives an indication of what some of those factors might be. We could take as our watchword the injunction of the Roman historian Titus Livius, better known to us as Livy. He explained that what makes a society strong is the well-being of its people — basic justice, basic opportunity, a modicum of spiritual reward — and the people's conviction that "the system" is set up to produce it. As Livy wrote, "An empire remains powerful so long as its subjects rejoice in it."

So here's the Titus Livius Hundred-Year Workout Plan:

First, instill an appreciation of the wider world. Start teaching it round instead of flat. Immigration helps us here. The influx of foreign students does too. And so does — seriously — America's entry, at last, into the world of soccer-playing countries. Colonial America defined itself as a nation as it advanced into an unknown interior; in a globalizing age the unknown world is as close to us as it was to any seventeenth-century settler. To drive home the idea that "we are not alone," there is no substitute for fluency in another language. Every educated person in the Roman Empire spoke at least two languages, and so did the strivers among the, uh, immigrant hordes. Americans have their priorities backward. They worry needlessly about the second part: whether the immigrants will ever learn English. They should be worrying about the first part: whether the elites will ever speak anything else.

Second, stop treating government as a necessary evil, and instead rely on it proudly for the big things it can do well. Privatization has its uses, and farming out government functions has its place — but the loss of civic engagement and loyalty across the board is a very real threat. The idea of doing for ourselves, of self-reliance, is part of the American myth; but letting government step in (to open the West, distribute land, nurture business, reduce poverty) is part of the American reality. The Social Security check every month, the safe drugs and highways, the guaranteed student loans, the health-care safety net in old age, the sandbags when the rivers flood — their inherent benefits aside, these things promote a sense of common alliance and mutual obligation that dwarf narrow considerations of "efficiency." They serve as a counterforce to inequality and the widening divisions of income and class. Besides, government can be held accountable in ways that the private sector can't. Yes, it takes some imagination to see how corrosive privatized government will prove to be many decades down the road — and that's another thing: start thinking in centuries.

Third, fortify the institutions that promote assimilation. "Empire-builders yearn for stability," writes Charles Maier, "but what imperial systems find hard to stabilize is, precisely, their frontiers." We can't change how the world works, can't change the laws of economics, can't move Mexico somewhere else, can't seal our border, and can't turn other countries into Shangri-la so that their people will stay home. Arrayed against all this, on high ground, is America's powerfully absorptive and transformative domestic culture. It's more than a match for any challenge, and doesn't need to be "run" by bureaucrats or told what to do. It just needs to be reinforced rather than undermined. In Massachusetts recently a debate broke out over whether the children of illegal aliens should be allowed to pay the low "in-state" rate at public colleges and universities. The answer should have been yes. The answer to public schools and public health services for immigrants should always be yes — these do more, with less, than any fence can accomplish. And yes should be the answer to a program of national service for all young people, which would revive the militia ethic of long ago. "We're all in it together" is a spirit that Rome lost. Nothing says "uan naishion, indivisibol" like national service.

Fourth, take some weight off the military. National service will help, at the margins. But if America enjoys the kind of economic vitality it hopes for, we're never going to attract enough qualified people into an all-volunteer military to perform all the global tasks we have in mind (and continue to dream up). No one wants to pay for an army of that size anyway. The alternative is to look at the demand side rather than the supply side, eliminating some of the things we need an army for. Rome, in the end, didn't have this option. As Edward Luttwak writes, it turned itself into a state whose entire purpose was military, and "ruthlessly extracted the food, fodder, clothing, arms, and money needed for imperial defense from an empire which became one vast logistic base." America is still free to make choices. It can let regional powers shoulder more of the burden. One unilateral decision above all would buy a lot of breathing room, and ought to be made regardless: to adopt a long-range energy policy, based as much as possible on renewable resources, allowing us to pull away, eventually, from military oversight of the entire Middle East. This may be a hundred-year project, but again, a society with pretensions to staying power thinks in those terms. Rome wasn't built in a day.

Whatever its specific fortunes, America will evolve into something very different from what it evolved out of, as the future will see more clearly than we can. For myself, I hope Americans manage to keep — and export — their egalitarianism, their entrepreneurship, and their exuberant impulse to associate in civic groups; and maybe manage to lose some of their hyper-individualism and their moralizing messianic streak. Looking ahead, I'm not sure America possesses one quality that Rome had in abundance: the stubborn urge, the absolute need, to persevere — to prevail at all costs in any undertaking, whatever the moral and human price might be.

But America does possess one quality that Rome didn't have at all. Rome's elites were deeply satisfied with their lot, their station, their state of mind. Their motto might have been *"Nihil potest ultra"* — a phrase from Cicero that Madison Avenue would render as "It doesn't get any better than this." Americans would glare in disbelief at Rome's self-satisfaction. Striving to make life "better than this," for ourselves and for others, for people living now and for those to come, is part of our social compact. Americans maintain a deep faith in the promise of invention and reinvention; they do not fear these things. Rome's economy was the same at the beginning as it was at the end: agrarian, Iron Age, preindustrial. America has lived through more social transformation in a few centuries than Rome did in a millennium. In less than two hundred years America has experienced the end of slavery, the leap from a farm to a high-tech economy, and an influx of alien newcomers whose presence, in percentage terms, is greater in size and proportion than the barbarian influx into Roman lands. We don't live in Mr. Jefferson's republic anymore, or in Mr. Lincoln's, or even in Mr. Eisenhower's. In America as in Rome, especially in disordered times, there is always the threat of cynicism. "We have long since lost the true name for things," Cato once lamented. But ask the question "What do you have faith in?" and ordinary Americans will give an answer that even the most privileged of Romans would not have: that improvement is possible. Improvement, in fact, is the point. The genius of America may be that it has built "the fall of Rome" into its very makeup: it is very consciously a constant work in progress, designed to accommodate and build on revolutionary change. Rome dissolved into history, successfully but only once. America has done so again and again.

Are we Rome? In important ways we just might be. In important ways we're clearly making some of the same mistakes. But the antidote is everywhere. The antidote is being American.

Acknowledgments

Ancient Rome is a vast, complicated, and slippery subject, and obviously so is modern America; even trained professionals approach with wary respect. I am deeply grateful to the many scholars and experts — on Rome and other topics — who offered advice and criticism, and words of encouragement and caution. They include C. R. Whittaker, at Cambridge University; Robin Birley and Barbara Birley, at the Vindolanda Trust, in Northumberland; Geoffrey S. Nathan, at the University of New South Wales; Ramsay MacMullen, at Yale University; Renate Kurzmann, at University College, Dublin; Cynthia Damon, Frederick Griffiths, and Anthony Marx, at Amherst College; Karen Barkey, at Columbia University; Shawn Graham, at the University of Manitoba; Stephen Fuller, at George Mason University; Charles Moskos, at Northwestern University; Gail Kern Paster, at the Folger Shakespeare Library; and Lawrence Korb, at the Center for American Progress. Conversations with many others in recent years influenced some of the thinking in this book, even before I decided to write it; I would mention in particular Richard Brookhiser, the biographer of George Washington and of Gouvernor Morris; Frederic Cheyette, at Amherst College; Francis Fukuyama, at the Johns Hopkins School of Advanced International Studies; Samuel Huntington, at Harvard University; the late Michael Kelly, of *The Atlantic Monthly*; Michael Lind, at the New America Foundation; Charles Maier, at Harvard University;

Jack Miles, at the Getty Research Institute; Thomas E. Ricks, at the *Washington Post;* Michael Sandel, at Harvard University; Joseph Stiglitz, at Columbia University; and Alan Wolfe, at Boston College.

Many friends and colleagues read versions of the manuscript, in whole or in part. For their comments and their time I am grateful to Graydon Carter, Paul Elie, James Fallows, David Friend, Robert D. Kaplan, Corby Kummer, Wayne Lawson, Toby Lester, Bernard-Henri Levy, Raphael Sagalyn, Eric Schlosser, Benjamin Schwarz, Martha Spaulding, Scott Stossel, Charles Trueheart, Robert Vare, and William Whitworth.

Much of the writing of this book was done at the Boston Athenaeum, on Beacon Hill. Richard Wendorf, the director; John Lannon, the deputy director; and the staff as a whole of that extraordinary library were of enormous help in countless ways. A number of people have assisted with research, most notably Emerson Hilton, and also Chris Berdik, Bessmarie Moll, Molly Finnegan, and Alyssa Rosenberg. The photographer Julian Cardona was my Virgil and companion in Ciudad Juarez, and I received guidance in Rome from Gerald O'Collins, S.J., of the Gregorian University, and Mario Marazitti, of the Community of San Egidio. Anton Mueller, at Houghton Mifflin, has been a sharp and patient counselor.

My wife, Anna Marie, has as usual been the shrewdest and most indefatigable editor of all.

Notes

page PROLOGUE: THE EAGLE IN THE MIRROR

1   "Urbs antiqua fuit": Virgil, *Aeneid*,1.12; 2.363.
road system is immense: Von Hagen, *Roads That Led to Rome*, p. 274; Adkins and Adkins, *Handbook to Life in Ancient Rome*, p. 172; Federal Highway Administration, U.S. Department of Transportation.
"peasants raced to report": The panegyrist Mamertinus, quoted in Williams, *Diocletian and the Roman Recovery*, p. 57.
will long be referred to: Lancon, *Rome in Late Antiquity*, pp. xii–xiii; Favro, *Urban Image of Augustan Rome*, pp. 116–117.

2   *the emperor's encampment*: Jones, *Later Roman Empire*, pp. 366–373, 636–640; Ferrill, *Fall of the Roman Empire*, p. 42.
20,000 of them: Carcopino, *Daily Life*, p. 70.
ministries of government: For details concerning a typical imperial entourage, see Jones, *Later Roman Empire*, pp. 366–373; Millar, *Emperor in the Roman World*, pp. 28–40.
bacon, cheese, and vinegar: Aelius Spartianus, "Hadrian," in Magie, trans., *Scriptores Historiae Augustae*, p. 31.
A letter survives: Casson, *Travel in the Ancient World*, p. 199.
the famous story: Danziger and Purcell, *Hadrian's Empire*, p. 135.

3   *Repairs to the Danube forts*: Williams, *Diocletian and the Roman Recovery*, p. 51.
"among the Quadi": Marcus Aurelius, *Meditations*, 1.18.
clustered around: Carcopino, *Daily Life*, pp. 70–71; Jones, *Later Roman Empire*, pp. 366–373.
eighteen-hour official visit: For security and other details of the presidential trip to Ireland, see Elisabeth Bumiller, "Bush Gets Chilly Reception on Eve of Meeting in Ireland," *New York Times*, June

26, 2004; Patrick Logue, "A Vision in Armour Plate," *Irish Times,* June 23, 2004; Angelique Chrisafis, "Irish Batten Down Hatches for Bush," *The Guardian,* June 25, 2004; Alison Hardie, "Irish Create a 'Ring of Steel' for Bush Visit," *The Scotsman,* June 26, 2004.

4 *Air Force One can carry:* Walsh, *Air Force One,* pp. 15–38; G. Robert Hillman, "Presidential Flight of Fancy," *Dallas Morning News,* February 20, 2005.

5 *allowed the United States:* Dan Bilefsky, "14 European Nations Actively Aided CIA Renditions, Report Asserts," *International Herald Tribune,* June 8, 2006.

   a senior British minister: Patten, *Cousins and Strangers,* p. 26.

6 *miniseries set in ancient Rome:* Alessandra Stanley, "HBO's Roman Holiday," *New York Times,* August 21, 2005; Nancy Franklin, "When in Rome," *The New Yorker,* July 4, 2005.

   Novels about Rome: Allan Massie, "Return of the Roman," *Prospect,* November 2006.

   Earlier films about Rome: William Fitzgerald, "Oppositions, Anxieties, and Ambiguities in the Toga Movie," in Joshel, Malamud, and McGuire, eds., *Imperial Projections,* pp. 23–49.

   cites Roman precedent: Nolan and Goyer, *Batman Begins,* p. 121.

   looking at their own reflections: Albert Schweitzer, *Quest of the Historical Jesus,* p. 4.

7 *"no mere international citizen":* Charles Krauthammer, "The Bush Doctrine," *Time,* March 5, 2001.

   "If people want to say": Quoted in James Atlas, "Leo-Cons: A Classicist's Legacy: New Empire Builders," *New York Times* ("Week in Review"), May 4, 2003.

   "jodhpurs and pith helmets": Max Boot, "The Case for an American Empire," *Weekly Standard,* October 15, 2001.

   "pointless . . . beyond challenge": George W. Bush, graduation speech at the United States Military Academy, June 1, 2002.

   "sorrows mounted up": Johnson, *Sorrows of Empire,* p. 285.

8 *"imperial overstretch":* Kennedy, *Rise and Fall of the Great Powers,* pp. 514–521.

   "empire in denial": Ferguson, *Colossus,* pp. 6–7, 28–29, 204, 301.

   "imperial understretch": Joseph S. Nye, "U.S. Power and Strategy After Iraq," *Foreign Affairs,* July–August 2003.

9 *lethal dynamic at work:* Jacobs, *Dark Age Ahead,* pp. 23–24.

   "The anti-Americans often invoke": Victor Davis Hanson, "I Love Iraq, Bomb Texas," *Commentary,* December 2002.

diverges from the message of Jesus: Horsley, *Jesus and Empire,* p. 6.

"God has raised up": Vinz, *Pulpit Politics,* p. 181.

Christmas card sent out: Associated Press, "Cheney Dodges Question of U.S. as Empire," January 24, 2004.

10 *"Who but America":* Debray, *Empire 2.0,* pp. 34–35

spread of televised contests: Pauline Arrillaga, "Meet the Champions of Chow," *Ottawa Citizen,* July 3, 2004.

11 *refers to the passageways:* See "vomitorium," *Oxford English Dictionary.*

pork-laden highway bill: Senator Trent Lott, "Rome's Roads," *U.S. Federal News,* May 13, 2005.

a speech from decades ago: Clare Boothe Luce, "Is the New Morality Destroying America?" *Human Life Review,* Summer 1978.

13 *a dozen case studies:* Neustadt and May, *Thinking in Time,* pp. 233–234.

"The only lesson": Quoted in Paul Johnson, "Where Hubris Came From," *New York Times Book Review,* October 23, 2005.

maintains a database: Eliot A. Cohen, "The Historical Mind and Military Strategy," *Orbis,* Fall 2005.

14 *little more than a theme park:* Habinek and Schiesaro, eds., *Roman Cultural Revolution,* p. 33.

in a famous lecture: Carl Becker, "Everyman His Own Historian," *American Historical Review* 37, no. 2 (1932), pp. 221–236.

"The simple natives": Tacitus, *Agricola,* 21.

15 *gargantuan statue:* John R. Patterson, "The City of Rome: From Republic to Empire," *Journal of Roman Studies* 82 (1992), pp. 186–215.

"Fewer have more": MacMullen, *Roman Social Relations,* p. 38.

words derived from Virgil: Virgil, *Eclogues,* 4.5.

17 *fully furnished frame of mind:* See, for instance, Gross, *Minutemen and Their World,* and Tillyard, *Elizabethan World Picture.*

20 *"The same strength":* Niebuhr, *Irony of American History,* p. 74.

21 *"In a world of ruins":* James, *Portrait of a Lady,* pp. 530–531.

Analysts of modern terrorism: Fred Kaplan, "Fighting Insurgents, by the Book," *Washington Post,* July 9, 2006.

Diocletian himself didn't see Rome: Williams, *Diocletian and the Roman Recovery,* pp. 187–188.

possible to visit Rome remotely: A reconstruction of portions of central Rome during antiquity can be found at two Web sites at the University of California at Los Angeles: www.cvrlab.org/projects/real_time/realtime_projects.html#ROME and dlib.etc.ucla.edu:8080/

projects/Forum. See also Lisa Guernsey, "Soaring Through Ancient Rome, Virtually," *Chronicle of Higher Education*, July 22, 2005, which describes efforts by Bernard Frischer, now at the University of Virginia, to make virtual fly-throughs of Rome more comprehensive, sophisticated, and accessible.

22 *like the guides and hawkers:* Casson, *Travel in the Ancient World*, pp. 262–291.

"Historians tell us": Freud, *Civilization and Its Discontents*, pp. 42–43.

23 *a psychic device:* Ibid., p. 44.

I. THE CAPITALS

24 *"Remember, Roman":* Virgil, *Aeneid*, 6.851–853.

"the indispensable nation": Quoted in Bob Herbert, "War Games," *New York Times*, February 22, 1998, among many other sources.

drew its last breath: Information about the events of 476 A.D. is drawn from Geoffrey Nathan, "The Last Emperor: The Fate of Romulus Augustulus," *Classica et Mediaevalia* 43 (1992), pp. 261–271, and Ralph W. Mathisen and Geoffrey Nathan, "Romulus Augustulus (475–476 A.D.) — Two Views," *De Imperatoribus Romanis: An Online Encyclopedia of Roman Emperors*. See also Bury, *Invasion of Europe*, pp. 166–183; Gibbon, *Decline and Fall*, vol. 2, pp. 321–347; and Heather, *Fall of the Roman Empire*, pp. 427–430.

25 *"I can neither forget nor express":* Gibbon, *Memoirs*, p. 134.

"Odoacer was the first barbarian": Gibbon, *Decline and Fall*, vol. 2, pp. 344–346.

26 *in many ways a sad and lonely one:* Martine Watson Brownley, "Gibbon: The Formation of Mind and Character," *Daedalus* 105, no. 3 (Summer 1976), pp. 13–25.

He observes gratuitously: Gibbon, *Decline and Fall*, vol. 2, p. 785 (fn. 63).

27 *"On Broad Potowmack's bank":* David Humphreys, "A Poem on the Future Glory of the United States of America," 1804. Quoted in Kenneth R. Bowling, "A Capital Before a Capitol," in Kennon, ed., *Republic for the Ages*, pp. 36–54.

settlement's early years: For some of Washington's Roman architectural antecedents, see Larry Van Dyne, "If These Stones Could Talk," *Washingtonian*, October 2002.

28 *"shall not wholly die":* Horace, *Odes*, 3.30, line 6.

reconstructing the landscape: See, for instance, Hare, *Walks in Rome*, passim.

29 *Several centuries as a monarchy:* The basic outline of Roman history is available in any number of excellent modern surveys. Two relied on here are Goodman, *Roman World,* and Potter, *Roman Empire at Bay.*

growth in maritime commerce: Keith Hopkins, "Taxes and Trade in the Roman Empire (200 B.C.–A.D. 400)," *Journal of Roman Studies* 70 (1980), pp. 101–125.

more than a million people: Lancon, *Rome in Late Antiquity,* p. 14; Carcopino, *Daily Life,* p. 65.

30 *Going back to those shipwrecks:* Keith Hopkins, "Taxes and Trade in the Roman Empire (200 B.C.–A.D. 400)," *Journal of Roman Studies* 70 (1980), pp. 101–125.

columns of colored marble: Dale Kinney, "Roman Architectural *Spolia,*" *Proceedings of the American Philosophical Society* 145, no. 2 (June 2001), pp. 138–150.

cattle were smaller: Ward-Perkins, *Fall of Rome,* p. 145.

A racist theoretician: Jasper Griffen, "Greeks, Romans, Jews & Others," *New York Review of Books,* March 16, 1989.

"Let students of Rome's decline": Kagan, *End of the Roman Empire,* p. viii.

In 1980, a German historian: The historian is Alexander Demandt. His 1984 book *Der Fall Roms* is cited in MacMullen, *Corruption,* p. ix; Ward-Perkins, *Fall of Rome,* p. 33; and elsewhere.

31 *"evils of a long peace":* Juvenal, *Satires,* 6.292.

"There will never be an end": Quoted in Jones, *Later Roman Empire,* p. 1025.

"defend the empire by praying for it": Origen, *Contra Celsum,* 8.73.

"too many idle mouths": Jones, *Later Roman Empire,* p. 1045.

32 *here's Nixon on the subject:* Quoted in Jasper Griffen, "Greeks, Romans, Jews & Others," *New York Review of Books,* March 16, 1989.

decline-of-Rome explanations: For concise summaries of the various theories about the Roman Empire's decline, see (among many possible sources) MacMullen, *Corruption,* pp. 1–57; Jones, *Later Roman Empire,* pp. 1025–1068; Heather, *Fall of the Roman Empire,* pp. 443–459; and Ferrill, *Fall of the Roman Empire,* pp. 10–22. Excerpts from the arguments of a range of historians are collected in Kagan, *End of the Roman Empire.*

33 *sack of the city by Alaric:* Ferrill, *Fall of the Roman Empire,* p. 103.

"The brightest light": Quoted in Ward-Perkins, *Fall of Rome,* p. 28.

took a toll on Rome's water supply: Lancon, *Rome in Late Antiquity,* pp. 14–15; Robert Coates-Stephens, "The Walls and Aqueducts of

Rome in the Early Middle Ages, A.D. 500–1000," *Journal of Roman Studies* 88 (1998), pp. 166–178; Llewellyn, *Rome in the Dark Ages*, pp. 78–108.

33 *began to plunder itself for* spolia: Dale Kinney, "Roman Architectural *Spolia*," *Proceedings of the American Philosophical Society* 145, no. 2 (June 2001), pp. 138–150. See also Joseph Alchermes, "*Spolia* in Roman Cities of the Late Empire," *Dumbarton Oaks Papers* 48 (1994), pp. 167–178; Beat Brenk, "*Spolia* from Constantine to Charlemagne: Aesthetics versus Ideology," *Dumbarton Oaks Papers* 41, (1987), pp. 103–109.

34 *The process of reversion:* The vegetal reclamation of some American inner cities has been chronicled in numerous press accounts. See, for instance, Naomi R. Patton, "Green Idea in Wayne County," *Detroit Free Press*, February 28, 2006; Bill McGraw, "Nature Restakes Its Claim as Trees Grow on Rooftops in Downtown Detroit," *Detroit Free Press*, May 17, 2001; Rochelle Riley, "Vacant Lot Blooms into Good Example," *Detroit Free Press*, July 21, 2003.

a sentence sagging with judgment and resignation: Gibbon, *Decline and Fall*, vol. 2, p. 438.

35 *"I shall scarcely give my consent":* Quoted in Craddock, *Edward Gibbon*, p. 114. See also Eliga H. Gould, "American Independence and Britain's Counter-Revolution," *Past and Present*, no. 154 (February 1997), pp. 107–141.

introduced to . . . Benjamin Franklin: Swain, *Edward Gibbon the Historian*, p. 83.

36 *steeped in the Roman code of* virtus: Richard Brookhiser, "A Man on Horseback," *Atlantic Monthly*, January 1996. Brookhiser's book *Founding Father* discusses George Washington and the colonial view of Rome at length. See also L. R. Lind, "Concept, Action, and Character: The Reasons for Rome's Greatness," *Transactions and Proceedings of the American Philological Association* 103 (1972), pp. 235–283.

as one historian sums it up: S. E. Smethurst, "The Growth of the Roman Legend," *Phoenix* 3, no. 1 (Spring 1949), pp. 1–14.

The Roman who epitomized republican ideals: Plutarch, *Lives*, vol. 8, pp. 237–411.

37 *stamped indelibly on the rhetoric:* Addison, *Cato*, 4.4; 2.4.

looked to preimperial Rome: Gummere, *American Colonial Mind*, pp. 173–190.

38 *epitome of America's Roman ideal:* Richard Brookhiser, "A Man on Horseback," *Atlantic Monthly*, January 1996.

obsessed with surveying: Dilke, *Greek & Roman Maps,* pp. 87–101.

an order from an English dealer: Gummere, *American Colonial Mind,* p. 15.

At Valley Forge: Brookhiser, *Founding Father,* p. 153.

39 *As the story goes:* Livy, *History of Rome,* 3.26.

"In two days": Quoted in Wills, *Cincinnatus,* p. 36.

a massive marble Washington: Ibid., pp. 67–72.

stock form of public presentation: Michael Lind, "The Second Fall of Rome," *Wilson Quarterly,* Winter 2000.

Thomas Cole's allegorical series: Stephen L. Dyson, "Rome in America," in Hingley, ed., *Images of Rome,* pp. 57–69.

40 *a pirate attack on Rome's port:* Robert Harris, "Pirates of the Mediterranean," *New York Times,* September 30, 2006.

postpone national elections: Michael Isikoff, "Election Day Worries," *Newsweek,* July 19, 2004. See also Ricardo Alonso-Zaldivar, "Few Will Discuss Postponing Vote if Terror Strikes," *Los Angeles Times,* July 22, 2004.

enter homes without knocking: David Stout, "Court Ruling Signals New Conservative Tilt," *International Herald Tribune,* June 16, 2006.

leaks of classified information: Adam Liptak, "Gonzales Says Prosecutions of Journalists Possible," *New York Times,* May 22, 2006; Stuart Taylor Jr., "Dumb and Dumber," *National Journal,* May 27, 2006.

"unitary executive theory": Christopher S. Kelley, "Rethinking Presidential Power — the Unitary Executive and the George W. Bush Presidency," paper presented at the annual meeting of the Midwest Political Science Association, April 2005. See also Paul Starobin, "Long Live the King!" *National Journal,* February 18, 2006, pp. 18–27; Christopher S. Yoo, Steven G. Calabresi, and Anthony Colangelo, "The Unitary Executive in the Modern Era, 1945–2004," *Iowa Law Review* 90, no. 2, pp. 601–731.

41 *the president has added signing statements:* Charlie Savage, "Cheney Aide Is Screening Legislation," *Boston Globe,* May 28, 2006; Jane Mayer, "The Hidden Power," *The New Yorker,* July 3, 2006.

couldn't get access to materials: Cassius Dio, *Roman History,* 53.19.3.

its undercover operatives: William G. Sinnigen, "The Roman Secret Service," *Classical Journal* 57, no. 2 (November 1961), pp. 65–72.

a vignette of entrapment: Epictetus, *Discourses,* 4.13.5.

a program known as Echelon: Susan Stellin, "Terror's Confounding Online Trail," *New York Times,* March 28, 2002. A number of Web sites purport to offer lists of key words the Echelon program listens for. See, for instance, www.ratbags.com/loon/2002/06echelon .htm and www.freerepublic.com/forum/a3b1842da7942.htm.

42 *Some of them . . . are self-explanatory:* Kieren McCarthy, "What Are Those Words That Trigger Echelon?" *The Register,* May 31, 2001.

which bore the name Carnivore: Susan Stellin, "Terror's Confounding Online Trail," *New York Times,* March 28, 2002.

sifts tens of millions: James Risen and Eric Lichtblau, "Bush Lets U.S. Spy on Callers Without Courts," *New York Times,* December 16, 2005; Leslie Cauley, "NSA Has Massive Database of Americans' Phone Calls," *USA Today,* May 11, 2006.

as James Madison observed: Madison, *Letters of Helvidius,* p. 36.

"The debates in the convention": Irons, *War Powers,* pp. 3–4, 11–27.

no president has ever sought: Leslie Gelb and Anne-Marie Slaughter, "Declare War," *Atlantic Monthly,* November 2005.

43 *"I'm the commander":* Bob Woodward, "A Course of 'Confident Action,'" *Washington Post,* November 19, 2002.

44 *originated in the study of old maps:* J. B. Harley, "Maps, Knowledge, and Power," in Cosgrove and Daniels, eds., *Iconography of Landscape,* pp. 277–312. For the application to Rome, see C. R. Whittaker, *Rome and Its Frontiers,* p. 78.

today would be considered "branding": William L. MacDonald, "Empire Imagery in Augustan Architecture," in *Archaeologia Transatlantica V* (1985). A full-length and meticulously illustrated treatment of the subject is Favro, *Urban Image of Augustan Rome.*

in the very center of Rome: Diane Favro, *"Pater urbis:* Augustus as City Father of Rome," *Journal of the Society of Architectural Historians* 51, no. 1 (March 1992), pp. 61–84.

a sundial the size of a football field . . . "The whole universe": John R. Patterson, "The City of Rome: From Republic to Empire," *Journal of Roman Studies* 82 (1992), pp. 186–215; Favro, *Urban Image of Augustan Rome,* pp. 260–264.

an oration in the Athenaeum: Aelius Aristides, "Regarding Rome," in Behr, *Aristides in Four Volumes,* vol. 2, pp. 11–13.

Vitruvius took up the same theme: Williams, *Romans and Barbarians,* p. 85.

45 *cartloads of garbage:* Favro, *Urban Image of Augustan Rome,* p. 255.

mountains of brick: Scholarship on the Roman brick industry is extensive. See Shawn Graham, *"Ex Figlinis:* The Network Dynamics of the Tiber Valley Brick Industry in the Hinterland of Rome," *Brit-*

ish Archaeological Reports, 2006; Shawn Graham, "Of Lumberjacks and Brick Stamps: Working with the Tiber as Infrastructure," in MacMahon and Price, eds., *Roman Working Lives,* pp. 106–124; Janet DeLaine, "The Baths of Caracalla: A Study in the Design, Construction, and Economics of Large-Scale Building Projects in Imperial Rome," *Journal of Roman Archaeology,* Supplementary Series 25 (1997); Meiggs, *Trees and Timber;* and Adam, *Roman Building Materials.*

single burn of a limestone kiln: Hughes, *Pan's Travail,* p. 126.

The docks at Ostia: Ibid., p. 84. See also Meiggs, *Roman Ostia,* pp. 8–10; plate 1.

The biggest component of the city's prodigious intake: For a comprehensive general description of the *annona* and its operations, see Rickman, *Corn Supply of Ancient Rome.*

46 *"The desired spectacle":* Gibbon, *Decline and Fall,* vol. 1, p. 81.

near the Porta Maggiore: Favro, *Urban Image of Augustan Rome,* p. 94.

work force required for the baking of bread: Jones, *Later Roman Empire,* pp. 699–701; Rickman, *Corn Supply of Ancient Rome,* pp. 204–209.

When Alaric laid siege to Rome: Ferrill, *Fall of the Roman Empire,* p. 103.

47 *they erected large public maps:* Williams, *Romans and Barbarians,* pp. 83–84. Dilke, *Greek & Roman Maps,* pp. 39–54.

At regular points along every roadway: Casson, *Travel in the Ancient World,* p. 173; Adkins and Adkins, *Handbook to Life in Ancient Rome,* pp. 182–183.

A monument in the capital: Everett L. Wheeler, "Methodological Limits and the Mirage of Roman Strategy: Part II," *Journal of Military History* 57, no. 2 (April 1993), pp. 215–240.

steady traffic in artwork: Casson, *Travel in the Ancient World,* pp. 247–251.

48 *replace the heads with their own:* Balsdon, *Romans & Aliens,* p. 178.

its Augustan phase began only in the twentieth century: Relevant studies of the creation of modern Washington include Abbott, *Political Terrain,* and Ricci, *Transformation of American Politics.* The source for economic and demographic data on the capital region is Stephen Fuller, director of the Center for Regional Analysis, George Mason University.

49 *"Only the residents of Washington":* Garreau, *Nine Nations of North America,* p. 101.

49   *the Great Game of foreign affairs and espionage:* Abbott, *Political Terrain,* p. 137.

50   *analysis of Washington phone books:* Ibid., p. 153.
     Ads for one Washington bank: "Background on Riggs Bank," washingtonpost.com, February 17, 2005.
     Washington's power-lunch restaurant: Tucker Carlson, "Power Host to Power Brokers in the Power Capital," *New York Times,* June 5, 2002.
     a four-part series: Allan Fotheringham, "Honeymoon May Be Over Before Clinton Says Vows," *Financial Post,* January 19, 1993.
     time obeys: Monte Reel, "Where Timing Truly Is Everything," *Washington Post,* July 22, 2003.
     "Once within the confines": Greenfield, *Washington,* pp. 28–29.

51   *new buildings behind old façades:* Paul L. Knox, "The Restless Urban Landscape: Economic and Cultural Change and the Transformation of Metropolitan Washington, D.C.," *Annals of the Association of American Geographers* 81, no. 2 (June 1991), pp. 181–209.
     Washington's wounded riposte: Shaw, *Caesar and Cleopatra,* p. 121.
     Petty's seventeenth-century Political Arithmetick: Alonso and Starr, *Politics of Numbers,* p. 14.
     program known as Total Information Awareness: Shane Harris, "TIA Lives On," *National Journal,* February 25, 2006.

52   *In locations throughout the city:* Anita Huslin, "If These Walls Could Talk. . . ," *Washington Post,* May 28, 2006.

53   *as much as 30 percent of the nightly network news . . . Of the 414 stories:* Andrew Tyndall, *Tyndall Report.* New York: ADT Research (www.tyndallreport.com/).
     "luxury-skybox view": James Wolcott, "Mighty Mouths," *The New Yorker,* December 26, 1993–January 2, 1994.
     the political week in Washington: James Fallows, "Did You Have a Good Week?" *Atlantic Monthly,* December 1994. See also James Fallows, "The New Celebrities in Washington," *New York Review of Books,* June 12, 1986.

54   *when Rome still had functioning electoral elements:* Andrew Lintott, "Electoral Bribery in the Roman Republic," *Journal of Roman Studies* 80 (1990), pp. 1–16.
     thought of Washington as an island: Abbott, *Political Terrain,* p. 4.
     Eisenhower complained: Ibid., p. 3.
     how isolating Washington is: Evan Thomas and Richard Wolffe, "Bush in the Bubble," *Newsweek,* December 19, 2005.

Dick Cheney's travel requirements: Elisabeth Bumiller, "Cheney's Needs on the Road: What, No NPR?" *New York Times*, March 23, 2006.

antibacterial gel: Mark Leibovich, "In Clean Politics, Flesh Is Pressed, Then Sanitized," *New York Times*, October 28, 2006.

55 *"the off-scourings of the city":* Livy, quoted in Emily Gowers, "The Anatomy of Rome from Capitol to Cloaca," *Journal of Roman Studies* 85 (1995), pp. 23–32.

the drains of the Colosseum: John R. Patterson, "The City of Rome: From Republic to Empire," *Journal of Roman Studies* 82 (1992), pp. 186–215.

"They contend with each other": Gibbon, *Decline and Fall*, vol. 2, p. 140. See also MacMullen, *Roman Social Relations*, p. 106.

the grandeur sought by . . . Kim Jong Il: Sang-Hun Choe, "North Korean Leader's Many Titles," Associated Press, March 8, 2004.

during the Kennedy administration: Cullen Murphy, "Feeling Entitled?" *Atlantic Monthly*, March 2005.

the quest for gloria: Cicero, *De Re Publica*, 5.7.

Romans spelled out their achievements: Robert K. Sherk, "Roman Geographical Exploration and Military Maps," *Aufstieg und Niedergang der romischen Welt* 2, no.1 (1974), pp. 534–562.

56 *a statement of good taste:* Anne Leen, "Cicero and the Rhetoric of Art," *American Journal of Philology* 112, no. 2 (Summer 1991), pp. 229–245.

"Its spectacular glass and columned façade": Paul L. Knox, "The Restless Urban Landscape: Economic and Cultural Change and the Transformation of Metropolitan Washington, D.C.," *Annals of the Association of American Geographers* 81, no. 2 (June 1991), pp. 181–209.

applause in the law courts: W. G. Runciman, "Capitalism Without Classes: The Case of Imperial Rome," *British Journal of Sociology* 34, no. 2 (June 1983), pp. 157–181.

an official triumph through the streets: Peter J. Holliday, "Roman Triumphal Painting: Its Function, Development, and Reception," *Art Bulletin* 79, no. 1 (March 1997), pp. 130–147.

quintessential Washington text: Ramsay MacMullen, "Roman Elite Motivation: Three Questions," *Past and Present* 88 (1980), pp. 8–16.

an "in-box imperium": Millar, *Emperor in the Roman World*, pp. 203–272.

57 *personally selected bombing targets:* Dallek, *Lyndon B. Johnson*, p. 223.

failed military rescue mission: Bowden, *Guests of the Ayatollah*, pp.

397–468; Richard Benedetto, "For the Most Part, Bush Lets Military Do Its Job," *USA Today*, January 23, 1991.

57   *vast political databases:* Todd S. Purdum, "Karl Rove's Split Personality," *Vanity Fair*, December 2006.

58   *resistance to multilateral arrangements:* A list of multilateral initiatives that the United States has either rejected or declined to comply with fully in recent years can be found at www.ieer.org/reports/treaties/factsht.html.

most recent federal budget: Thom Shanker, "In Bill's Fine Print, $20 Million to Celebrate Victory in the War," *New York Times*, October 4, 2006.

after the last emperor was deposed: Brown, *World of Late Antiquity*, p. 131.

2. THE LEGIONS

59   *"Valens was overjoyed":* Ammianus Marcellinus, *Roman History*, 31.4.6.

"I would go further": Max Boot, "Defend America, Become American," *Los Angeles Times*, June 16, 2005.

Bagram has yielded glassware: "A Virtual Catalogue of the Begram Ivory and Bone Carvings," Electronic Cultural Atlas (ecai.org/begramweb/), an initiative headed by Jeanette Zernenke. See in particular Sanjyot Mehendale, "Begram: New Perspectives on the Ivory and Bone Carvings," doctoral dissertation, University of California, Berkeley, 2005.

60   *Bagram today is an outpost:* For details about Bagram Air Base, see Kaplan, *Imperial Grunts*, pp. 196–202; Michael R. Gordon, "Securing Base, U.S. Makes Its Brawn Blend In," *New York Times*, December 3, 2001; Bruce Rolfsen, "Comfort Is a Relative Term in Afghanistan," *Air Force Times*, August 29, 2005; "Afghanistan — Bagram Airbase," GlobalSecurity.org.

naturalization ceremonies have been held: Tiffany Evans, "23 Servicemembers in Afghanistan Become U.S. Citizens," Armed Forces Press Service, July 29, 2005.

celebrities from the United States: Susan Dominus, "Not Bob Hope's U.S.O.," *New York Times Magazine*, November 13, 2005; Tamara Jones, "The U.S.O.'s Handshake Squad," *Washington Post*, December 24, 2005.

61 *A single company, Kellogg Brown & Root:* KBR's role is explained in numerous sources, including Web sites at the Pentagon (www .amc.army.mil/logcap/WhoWhere1.html) and the Center for Public Integrity (www.publicintegrity.org/wow/bio.aspx?act=pro &ddlC=31), and in a *Frontline* documentary (www.pbs.org/wgbh/pages/frontline/shows/warriors/view/).

more than a thousand fast-food restaurants: Testimony by Kathryn G. Frost, commander, Army and Air Force Exchange Service, before the House Armed Services Committee, Subcommittee on Military Personnel, April 7, 2005.

an American correspondent . . . writes: Kaplan, *Imperial Grunts,* p. 197.

62 *Vindolanda's moats and ditches:* For this and other general information about Vindolanda, see Bowman, *Life and Letters;* Birley, *Garrison Life.*

no idea what will come out of the ground: Interview with Robin Birley, Vindolanda, December 2005.

63 *they had proved unreliable:* Birley, *Garrison Life,* pp. 41–48.

Their bitter anthem: From Auden, "Twelve Songs," *Collected Poems,* p. 143.

Fly over the Syrian desert: Kennedy and Riley, *Rome's Desert Frontier,* passim.

to amuse the troops: Trajan's action is cited in C. E. Manning, "Acting and Nero's Conception of the Principate," *Greece & Rome* 22, no. 2 (October 1975), pp. 164–175.

64 *delicate bits of writing:* Bowman, *Life and Letters,* pp. 9–19; Birley, *Garrison Life,* pp. 15–40.

"Octavius to his brother": Birley, *Garrison Life,* pp. 114–116; Bowman, *Life and Letters,* pp. 136–137.

"Nails for boots": Cited in R.S.O. Tomlin, "The Vindolanda Writing Tablets," *Britannia* 27 (1996), pp. 459–463; Birley, *Garrison Life,* pp. 114, 85.

a Latin epithet: Birley, *Garrison Life,* p. 95.

"I am surprised": Ibid., p. 107.

"I am beginning to wonder": Allen Breed, "Internet Letters Tell of Couple's Heartache During the Iraq War," Associated Press, January 11, 2004.

"Be good for Grandma": Matt Richtel, "A Nation at War: Email," *New York Times,* March 23, 2003.

"Send me some cash": Bowman, *Life and Letters,* p. 138.

64 *"Still no hope in sight"*: Allen Breed, "Internet Letters Tell of Couple's Heartache During the Iraq War," Associated Press, January 11, 2004.

"While I am writing": Birley, *Garrison Life*, p. 118.

"I have sent you": Birley, *Roman Documents*, p. 19.

"For the celebration": Birley, *Garrison Life*, p. 136.

"Well, on that happy note": Evans, "Two Years in Iraq," *Newsday*, March 22, 2005.

"seemingly imperial power": Joseph Nye, *All Things Considered*, National Public Radio, March 11, 2002.

65 *when visiting American forces*: Quoted in Hugh Honour, "From Here to Eternity," *New York Review of Books*, June 13, 1991.

its level of annual spending: Robert A. Pape, "Soft Balancing Against the United States," *International Security*, Summer 2005; Robert S. Dudney, "What It Means to Be number 1," *Air Force*, February 2006; Fareed Zakaria, "Our Way," *The New Yorker*, October 14, 2002; David Brooks, "Why the U.S. Will Always Be Rich," *New York Times Magazine*, June 9, 2002; Emily S. Rosenberg, "Bursting America's Imperial Bubble," *Chronicle Review*, November 3, 2006.

we provide a security umbrella: John Ikenberry, "Illusions of Empire," *Foreign Affairs*, March/April 2004.

66 *had to squeeze its people hard*: Mattern, *Rome and the Enemy*, pp. 136–137; Jones, *Later Roman Empire*, p. 1039; Luttwak, *Grand Strategy of the Roman Empire*, p. 130.

Perhaps the next stop: William G. Sinnigen, "The Roman Secret Service," *Classical Journal* 57, no. 2 (November 1961), pp. 65–72; P. K. Baillie Reynolds, "The Troops Quartered in the *Castra Peregrinorum*," *Journal of Roman Studies* 13 (1923), pp. 168–189.

67 *one of the imperial stud farms*: Davies, *Service in the Roman Army*, pp. 163, 167.

Next: a tour of the fabricae: Simon James, "The *Fabricae*: State Arms Factories of the Later Roman Empire," in Coulston, ed., *Military Equipment and the Identity of Roman Soldiers*, pp. 257–331.

than Rome usually possessed: Luttwak, *Grand Strategy*, p. 2; Maier, *Among Empires*, p. 71.

would rank No. 50: Defense Logistics Agency, official history (www.dla.mil/history/history.htm).

68 *In one recent year*: Communication from Diana Stewart, public-affairs officer, Defense Supply Center, April 2006.

a thing of organizational beauty: For an extended account, see William Langewiesche, "Peace Is Hell," *Atlantic Monthly,* October 2001.

turns impressive feats: U.S. Army Special Operations Command; Ann Scott Tyson, "Inside the Army's Quest for Agility, Speed," *Christian Science Monitor,* October 2, 2002.

the globe-circling Royal Navy: Rodger, *Command of the Ocean,* pp. 306–307.

a single Roman legion . . . required: Goldsworthy, *Roman Army at War,* p. 290.

criticized for its enormous fuel consumption: Robert Bryce, "Gas Pains," *Atlantic Monthly,* May 2005.

In meat alone: Goldsworthy, *Roman Army at War,* pp. 292–296.

a sickle was standard: Ibid., p. 291.

steady stream of supplies: Michael Fulford, "Territorial Expansion and the Roman Empire," *World Archaeology* 23, no. 3 (February 1992), pp. 294–305; José Remesal Rodriguez, "Baetica and Germania: Notes on the Concept of 'Provincial Interdependence' in the Roman Empire," in Erdkamp, *Roman Army and the Economy,* pp. 293–308.

69 *Excavated Roman latrines:* B. A. Knights, Camilla A. Dickson, J. H. Dickson, and D. J. Breeze, "Evidence Concerning the Roman Military Diet at Bearsdon, Scotland, in the 2nd Century AD," *Journal of Archaeological Science* 10 (1983), pp. 139–152; Whittaker, *Frontiers of the Roman Empire,* p. 103; Elton, *Frontiers of the Roman Empire,* p. 81.

Caesar . . . built a wooden bridge: Caesar, *Gallic War,* 4.17–19.

a new riverine fleet: Williams, *Romans and Barbarians,* p. 104.

American standards of readiness: James Fallows, "Why Iraq Has No Army," *Atlantic Monthly,* December 2005.

military handbook from the second century: Davies, *Service in the Roman Army,* p. 103.

70 *an assessment of Roman conditioning:* Keppie, *Making of the Roman Army,* p. 198.

tightly knit and self-contained: Ramsay MacMullen, "The Legion as a Society," *Historia* 33 (1984), pp. 440–456.

"All professional armies": Goldsworthy, *Roman Army at War,* p. 252.

Roman garrisons were much the same: Ramsay MacMullen, "The Legion as a Society," *Historia* 33, no. 4 (1984), pp. 440–456.

Roman tomb inscription: Simon James, in Coulston, ed., *Military Equipment,* p. 281.

the equivalent of Point Blank Body Armor: This is one of several companies that supply the U.S. military with body armor. See DHB Industries Inc. (www.dhbt.com). Also Michael Moss, "Many Missteps Tied to Delay of Armor to Protect Troops," *New York Times,* March 7, 2005; Michael Moss, "U.S. Struggling to Get Soldiers Updated Armor," *New York Times,* August 14, 2005.

71 *"imperial overstretch":* The classic formulation is in Kennedy, *Rise and Fall of the Great Powers.* For the specific application to the United States, see pp. 514–535.

his influential monograph: Luttwak, *Grand Strategy of the Roman Empire,* p. 1. An excellent concise overview of many of the same issues is provided by Ferrill, *Roman Imperial Grand Strategy.*

taken to task: Eliot Cohen, review of *The Grand Strategy of the Roman Empire,* by Edward Luttwak, *Political Science Quarterly* 93, no. 1 (Spring 1978), pp. 174–175.

great stock in battlefield ferocity: Mattern, *Rome and the Enemy,* pp. 171–174.

the influence of "soft" power: The concept is discussed at length in Nye, *Paradox of American Power,* pp. 8–12.

72 *seating him directly under a U.S. cruise missile:* Holbrooke, *To End a War,* pp. 244–245.

to encourage a spirit of mature reflection: Cassius Dio, *Roman History,* 69.9.6

"proceeding . . . along a path": Tocqueville, *Democracy in America,* p. 434.

these ideas no longer have significant competition: His views were first expressed in a 1989 article in the journal *The National Interest* but received their fullest expression in Fukuyama, *End of History.*

equal measures of Ronald Reagan and Emma Lazarus: Cited in Williams, *Romans and Barbarians,* p. 85.

73 *The larger goal of American foreign policy:* Benjamin Schwarz, "Why America Thinks It Has to Run the World," *Atlantic Monthly,* June 1996.

"consensual empire": Charles S. Maier, "An American Empire?" *Harvard,* November/December 2002.

74 *"eggshell strategy":* Ferrill, *Roman Imperial Grand Strategy,* p. 35.

about twenty-five emperors: Ibid., p. 43.

The emperor Valerian: Potter, *The Roman Empire at Bay,* pp. 254–256; Gibbon, *Decline and Fall,* vol. 1, pp. 234–237.

"our present revenues": Cassius Dio, *Roman History,* 52.6.

75   *"The machinery of empire"*: Luttwak, *Grand Strategy*, p. 5.

three quarters of the . . . research money . . . Half of America's scientists: Barney Warf and Amy Glasmeier, "Military Spending, the American Economy, and the End of the Cold War," *Economic Geography* 69, no. 2 (April 1993), pp. 103–106. See also James Galbraith, "The Unbearable Costs of Empire," *American Prospect,* November 18, 2002.

"Every gun that is made": Quoted in Barney Warf and Amy Glasmeier, "Military Spending, the American Economy, and the End of the Cold War," *Economic Geography* 69, no. 2 (April 1993), pp. 103–106.

it maintains basing rights: Johnson, *Sorrows of Empire*, p. 4; Kaplan, *Imperial Grunts*, p. 7.

76   *three hundred military sites:* They are listed and described in Bidwell, *Roman Forts in Britain*.

a size not reached again: Ferrill, *Roman Imperial Grand Strategy*, p. 1.

77   *at one point during the reign of Tiberius:* Tacitus, *Annals*, 4.5.

Roman legions anticipated this practice: Keppie, *Making of the Roman Army*, pp. 205–212.

"Given the worldwide array": Kennedy, *Rise and Fall of the Great Powers*, p. 529.

too large to be affordable: Walter Goffart, "Rome, Constantinople, and the Barbarians," *American Historical Review* 86, no. 2 (April 1981), pp. 275–306. See also Jones, *Later Roman Empire*, p. 1035.

78   *a concerned Roman citizen:* For biographical inferences and historical context, see E. A. Thompson, trans., *Roman Reformer*, pp. 1–6. See also the introduction by M.W.C. Hassall, in Hassall and Ireland, eds., *De Rebus Bellicis*.

labor shortages are a problem: Jones, *Later Roman Empire*, p. 1042.

79   *which propel the ship:* Thompson, *Roman Reformer*, p. 119.

"technologically as backward": Jones, *Later Roman Empire*, p. 1047.

Measuring distances: Robert K. Sherk, "Roman Geographical Exploration and Military Maps," *Aufstieg und Niedergang der romanischen Welt*, pp. 534–562.

technology of waterpower: Jones, *Later Roman Empire*, p. 1048.

Germans today are number 3: Richard Florida, "The World Is Spiky," *Atlantic Monthly*, October 2005.

the barbarians, not the Romans: Lynn White Jr., "Technology and Invention in the Middle Ages," *Speculum* 15, no. 2 (April 1940), pp. 141–159.

80   *wouldn't have solved the . . . problem:* Ferrill, *Roman Imperial Grand Strategy*, p. 59.

one barbarian recruit: Elton, *Frontiers of the Roman Empire*, p. 65.

advancement . . . was rebuffed: Heather, *Fall of the Roman Empire*, pp. 216–226.

"he and his men": MacMullen, *Corruption*, p. 204.

81   *number of soldiers with roots in Italy:* Webster, *Roman Imperial Army*, pp. 107–109.

but after Ammianus Marcellinus: Burns, *Barbarians Within the Gates*, p. xvi.

"a society rather sealed off": Ramsay MacMullen, "The Legion as a Society," *Historia* 33, no. 4 (1984), pp. 440–456.

In his 1957 study: Huntington, *Soldier and the State*, pp. 143–157, 456–466.

Princeton graduates in the class of 1956: Data provided by the military sociologist Charles Moskos, of Northwestern University, April 2006.

82   *At a recent convention of recruiters:* Ibid.

Only in the military do white people: Ibid.

83   *"It is no longer enough":* Quoted in Thomas E. Ricks, "The Widening Gap Between the Military and Society," *Atlantic Monthly*, July 1997.

the Roman army grew: Luttwak, *Grand Strategy*, p. 177; Ferrill, *Fall of the Roman Empire*, pp. 41–43.

One emperor ordered the thumbless ones: Ferrill, *Roman Imperial Grand Strategy*, pp. 46–47.

"avoiding having to face two major military threats": Luttwak, *Grand Strategy*, p. 152. A. R. Hands, "The Fall of the Roman Empire in the West: A Case of Suicide or 'Force Majeure'?" *Greece & Rome* 10, no. 2 (October 1963), pp. 153–168.

84   *America no longer possesses:* Thom Shanker and Eric Schmitt, "Pentagon Weighs Strategy Change to Deter Terror," *New York Times*, July 5, 2005.

hemorrhaging young officers: Thom Shanker, "Young Officers Leaving Army at a High Rate," *New York Times*, April 10, 2006.

severe recruiting shortfalls: William J. Perry, chair, National Security Advisory Group, *The U.S. Military: Under Strain*, January 2006.

the pace of rotation: Interview with Thomas E. Ricks, military correspondent for the *Washington Post*, April 2004. See also Andrew F. Krepinevich, "The Thin Green Line," Center for Strategic and Budgetary Assessments, August 2004; James Fallows, "Why Iraq Has No Army," *Atlantic Monthly*, December 2005.

"a race against time": Andrew F. Krepinevich, "The Thin Green Line," Center for Strategic and Budgetary Assessments, August 2004. See also Lawrence J. Korb, "All-Volunteer Army Shows Signs of Wear," *Atlanta Journal-Constitution*, February 27, 2005; Lawrence J. Korb, "How to Update the Army's Reserves," *Foreign Affairs*, March/April 2004; Korb, *Reshaping America's Military*.

modern military classic: Fehrenbach, *This Kind of War*, p. 290.

an experiment called Project 100,000: Kelly M. Greenhill, "Don't Dumb Down the Army," *New York Times*, February 17, 2006.

85   *the Pentagon announced last year*: Michael Kilian and Deborah Horan, "Enlistment Drought Spurs New Strategies," *Chicago Tribune*, March 31, 2005; Army News Service, June 22, 2006.

To lower the bar even further: "The Army Wants You — Even with Tattoos," *Washington Post*, April 3, 2006; Douglas Belkin, "Struggling for Recruits, Army Relaxes Its Rules," *Boston Globe*, February 20, 2006.

"white man's burden": Ferguson, *Colossus*, p. 295.

least qualified in a decade: Fred Kaplan, "The Dumbing Down of the U.S. Army," *Slate*, October 4, 2005.

86   *ordered commanders to reduce high attrition*: The memo was published in *Slate*. See Phillip Carter and Owen West, "Dismissed!" *Slate*, June 2, 2005.

"From Pago Pago": James Brooke, "On Farthest U.S. Shores, Iraq Is a Way to a Dream," *New York Times*, July 31, 2005.

Roman citizenship after twenty-five years: Ferrill, *Roman Imperial Grand Strategy*, p. 5.

"We could model a Freedom Legion": Max Boot, "Defend America, Become American," *Los Angeles Times*, June 16, 2005.

cobbled together an ad hoc Coalition of the Willing: GlobalSecurity.org.

2,000 monkeys: Dana Milbank, "Many Are Willing but Few Are Able," *Washington Post*, March 25, 2003.

87   *"privatized military industry"*: Singer, *Corporate Warrior*, pp. 14–16. See also Avant, *Market for Force*, pp. 113–138.

Security at the U.S. Military Academy: Marek Fuchs, "Out of Khaki, into Blue at the Gates of West Point," *New York Times*, January 22, 2004; Leslie Wayne, "Security for Homeland, Made in Alaska," *New York Times*, August 12, 2004.

87 *A guard at the gate:* Interview with James Fallows, April 2006.
one component of an international industry: Singer, *Corporate Warriors,* pp. 3–18.
Milo Minderbinder's remark: Heller, *Catch-22,* p. 269.
gone to private contractors: The amount of public money awarded to private contractors in Iraq may never be known with any certainty. The agency that exercises oversight over spending by private contractors in Iraq, the Office of the Special Inspector General for Iraq Reconstruction, has been ordered to shut down operations by October 1, 2007, under the terms of legislation signed into law by President Bush in October 2006.

88 *the number excludes another 100,000:* Renae Merle, "Census Counts 100,000 Contractors in Iraq," *Washington Post,* December 5, 2006.
Others, like Triple Canopy: Daniel Bergner, "The Other Army," *New York Times Magazine,* August 14, 2005.

89 *Vegetius described the real-world calamities:* Ferrill, *Roman Imperial Grand Strategy,* p. 51.
raises all the same concerns: Andrew F. Krepinevich, "The Thin Green Line," Center for Strategic and Budgetary Assessments, August 2004.
a former counterterrorism chief . . . addressed a meeting: Kelly Kennedy, "Firm Offers Itself as Army for Hire," *Army Times,* April 10, 2006.

3. THE FIXERS

91 *"Four or five men get together":* Quoted in MacMullen, *Corruption,* p. 147.
"Corruption charges": Syriana. Screenplay by Steven Gaghan; based on the book *See No Evil,* by Robert Baer. Warner Brothers Pictures, 2005, p. 93.
old medical museum: The museum is located at the Ospedale di Santo Spirito in Sassia, in the Borgo Santo Spirito. (The words "in Sassia" refer to the fact that the original structure, built in the eighth century, catered to Saxon pilgrims in Rome.)

92 *Pliny's life was enviable:* Sherwin-White, *Letters of Pliny,* pp. 69–82. See also Fred S. Dunham, "The Younger Pliny," *Classical Journal* 40, no. 7 (April 1945), pp. 417–426; Graves H. Thompson, "Pliny's 'Want of Humor,'" *Classical Journal* 37, no. 4 (January 1942), pp. 201–209;

and John Newbold Hough, "A Few Inefficiencies in Roman Provincial Administration," *Classical Journal* 35, no. 1 (October 1939), pp. 17–26.

one kind of letter stands out: Pliny, *Letters*, vol. 2, pp. 183, 197, 207, 311, 313.

93   *"When I was seriously ill":* Ibid., p. 173.

"Your kindness, noble emperor": Ibid., p. 171.

94   *A notable proportion:* G.E.M. De Ste. Croix, "*Suffragium:* From Vote to Patronage," *British Journal of Sociology* 5, no. 1 (March 1954), pp. 33–48.

The very term "letters of recommendation": Hannah M. Cotton, "*Mirificum Genus Commendationis:* Cicero and the Latin Letter of Recommendation," *American Journal of Philology* 106, no. 3 (Autumn 1985), pp. 328–334.

the consummate courtier: C. R. Haines, "Fronto," *Classical Review* 34, no. 1–2 (February–March, 1920), pp. 14–18.

That his letters . . . survive at all: Champlin, *Fronto*, p. 3.

glimpses of daily life: Haines, *Correspondence of Marcus Cornelius Fronto*, p. 15.

95   *as the editor . . . points out:* C. R. Haines, "Fronto," *Classical Review* 34, no.1–2 (February–March 1920), pp. 14–18.

"If you love me": Fronto, *Correspondence*, vol. 2, p. 175.

"The bearer, his relative and namesake": Libanius, *Autobiography*, vol. 2, p. 19.

96   *"that cloud of Zeus":* Scott Bradbury, "A Sophistic Prefect: Anatolius of Berytus in the Letters of Libanius," *Classical Philology* 95, no. 2 (April 2000), pp. 172–186.

the history of a single Latin term: G.E.M. de Ste. Croix, "*Suffragium:* From Vote to Patronage," *British Journal of Sociology* 5, no. 1 (March 1954), pp. 33–48. See also Jones, *Later Roman Empire*, pp. 391–394.

98   *"the diverting of governmental force":* MacMullen, *Corruption*, p. 123.

99   *"From the parasite do-nothing":* Carcopino, *Daily Life*, p. 171.

100   *stock targets:* Damon, *Mask of the Parasite*, provides numerous examples; see, for instance, pp. 105–191.

its way of conducting foreign affairs: Cunliffe, *Greeks, Romans & Barbarians*, pp. 177–184; Lintott, *Imperium Romanum*, pp. 168–174.

The longtime Saudi ambassador: Robert Baer, "The Fall of the House of Saud," *Atlantic Monthly*, May 2003.

class stratification . . . was extreme: MacMullen, *Roman Social Relations*, pp. 88–89.

100   *the nation's worsening income inequality:* Isaac Shapiro, "New IRS
      Data Show Income Inequality Is Again on the Rise," Center on
      Budget and Policy Priorities, October 17, 2005.
      The latest data: Steven Greenhouse and David Leonhardt, "Real
      Wages Fail to Match a Rise in Productivity," *New York Times,* Au-
      gust 28, 2006.
      on the order of 5,000 or 10,000 to one: Schiavone, *End of the Past,* p. 71.
101   *the sumptuous* horti: John R. Patterson, "The City of Rome: From
      Republic to Empire," *Journal of Roman Studies* 82 (1992), pp. 186–215;
      Favro, Urban Image of Augustan Rome, pp. 177–180.
      Agrippa in a single year: Cassius Dio, *Roman History,* 49.43.
      a man of real substance: Sogno, *Q. Aurelius Symmachus,* passim; Bar-
      row, *Prefect and Emperor,* p. 14; J. A. McGeachy Jr., "The Editing of
      the Letters of Symmachus," *Classical Philology* 44, no. 4 (October
      1949), pp. 222–229.
      Elsewhere he complains: Heather, *Fall of the Roman Empire,* p. 21.
      Symmachus notes that twenty-nine Saxon captives: J. A. McGeachy, Jr.,
      "The Editing of the Letters of Symmachus," *Classical Philology* 44,
      no. 4 (October 1949), pp. 222–229.
102   *O'Hara's fictional Gibbsville:* Benjamin and Christina Schwarz, "John
      O'Hara's Protectorate," *Atlantic Monthly,* March 2000. On the im-
      portance of the curial class, see MacMullen, *Corruption,* p. 44.
      "did not work for the common benefit": Diane Favro, " 'Pater Urbis': Au-
      gustus as City Father of Rome," *Journal of the Society of Architectural
      Historians* 51, no. 1 (March 1992), pp. 61–84.
103   *"genial, oily":* MacMullen, *Corruption,* p. 126.
      "Thank you for those fieldfares": Ibid., p. 126.
      "an arbata of olives": Ibid.
      "The Gentleman who will have the Honour": Willcox, ed., *Papers of
      Benjamin Franklin,* pp. 499–500.
104   *"This political operation":* Andrew Ferguson, "A Lobbyist's Prog-
      ress," *Weekly Standard,* December 20, 2004. A complete compila-
      tion of the Jack Abramoff e-mails, released by the Senate Indian
      Affairs Committee, can be found at indian.senate.gov/public/
      index.cfm?FuseAction=Hearings.Investigations.
      "Our friend . . . asked": Ferguson, "A Lobbyist's Progress," *Weekly
      Standard,* December 20, 2004.
      "Da man!": Michael Crowley, "A Lobbyist in Full," *New York Times
      Magazine,* May 1, 2005.
      What accounts for this change?: MacMullen, *Corruption,* pp. 122–170.

104   *A bronze plaque:* Ibid., p. 151.

105   *the Republican contribution hierarchy:* Thomas B. Edsall, Sarah Cohen, and James V. Grimaldi, "Pioneers Fill War Chest, Then Capitalize," *Washington Post,* May 16, 2004.

emperors are standing athwart the tide: MacMullen, *Corruption,* pp. 149–156.

"Orders have to be followed": Ibid., p. 172.

106   *"His abilities were not inadequate":* Gibbon, *Decline and Fall,* vol. 1, p. 881. Other accounts of the Lepcis episode can be found in Heather, *Fall of the Roman Empire,* pp. 100–104; MacMullen, *Corruption,* pp. 154–155, 179–180, 193–194; and Ammianus Marcellinus, *Roman History,* 28.6.

107   *"the rigid impartiality of Palladius":* Gibbon, *Decline and Fall,* vol. 1, p. 881.

108   *made the military no larger:* MacMullen, *Corruption,* pp. 172–177, 183–185.

The precise definition of "feudalism": F. L. Ganshof, *Feudalism,* p. xv. For a recent perspective on medieval tendencies in the modern world, see John Rapley, "The New Middle Ages," *Foreign Affairs* (May–June 2006).

109   *still includes influence peddling and bribery:* "Talking Points Memo Document Collection," February 18, 2006 (www.talkingpoints memo.com/docs/cunningham/); Thomas Frank, "What Is K Street's Project?" *New York Times,* August 19, 2006.

incompetent cronies: "Welcome to the Hackocracy," *New Republic,* October 17, 2005.

110   *Privatization along these lines:* Guttman and Willner, *Shadow Government,* p. xiii.

111   *a presidential commission on privatization:* Linowes, *Privatization,* p. 1.

"We would do well": Harlan Cleveland, quoted in Osborne and Gaebler, *Reinventing Government,* pp. 39–40.

crossing from the temporal realm: Saritha Rai, "Short on Priests, U.S. Catholics Outsource Prayers to Indian Clergy," *New York Times,* June 13, 2004.

"While all professions": Huntington, *Soldier and the State,* pp. 14–15.

112   *virtually nonexistent:* Ramsay MacMullen, "Personal Power in the Roman Empire," *American Journal of Philology* 107, no. 4 (Winter 1986), 512–524.

Private security is a major growth industry: Paul Parfomak, *Guarding America: Security Guards and U.S. Critical Infrastructure Protection,*

Congressional Research Service, November 12, 2004; Clifford D. Shearing and Philip C. Stenning, "Private Security: Issues for Social Control," *Social Problems* 30, no. 5 (June 1983), pp. 493–506.

112 *buy themselves some private arbitration:* Editorial, "'Rent-a-Judge' Good for Taxpayers," *Rocky Mountain News,* July 28, 2005; Martin Kasindorf, "Rent-a-Judges Forced Out of California Courts," *USA Today,* April 25, 2003; Ted Gest and Steve L. Hawkins, "Slug It Out in the Privacy of Your Own Courtroom," *U.S. News & World Report,* December 5, 1988; Richard Lacayo, "Tell It to the Rent-a-Judge," *Time,* August 29, 1988.

turn the prison industry over to the private sector: Eric Schlosser, "The Prison-Industrial Complex," *Atlantic Monthly,* December 1998.

Entrepreneurs known as "bed-brokers": Ibid.; Rena Singer, "View and Space Lousy, but the Realty Sells," *Philadelphia Inquirer,* September 7, 1997; Robert Bryce, "Texas Locks Up Market in Spare Jail Beds," *Los Angeles Times,* August 13, 1995.

In Rome, private companies sprang up: Schiavone, *End of the Past,* pp. 121–122.

113 *state support is diminishing:* Sam Dillon, "At Public Universities, Warnings of Privatization," *New York Times,* October 16, 2005.

"slow slide": Ibid.

BankAmerica Dean: Washburn, *University, Inc.,* p. 5.

Most of the funding: Ibid., p. xv.

Ken Lay Center: Ibid., p. xv.

114 *"Kmart's attitude has always been":* Ibid., p. 5.

Meat inspection is done largely: Melody Petersen and Christopher Drew, "The Slaughterhouse Gamble," *New York Times,* October 10, 2003.

Americans were up in arms: Simon Romero and Heather Timmons, "A Ship Already Sailed," *New York Times,* February 24, 2006.

proposals to privatize Social Security: Joseph E. Stiglitz, "Progressive Dementia," *Atlantic Monthly,* November 2005.

the new Medicare prescription-drug plan: Milt Freudenheim, "Medicare Drug Plan Feeds More Profitable Managed Care," *New York Times,* March 31, 2006; Milt Freudenheim, "A Windfall from Shifts to Medicare," *New York Times,* July 18, 2006; Alan Sager and Deborah Socollar, "How Much Would Drug Makers' Profits Rise Under a Medicare Prescription Drug Benefit?" Boston University School of Public Health, April 2004.

government by user fee: Paul Starr and Ross Corson, "Who Will Have the Numbers? The Rise of the Statistical Services Industry and the Politics of Public Data," in Alonso and Starr, *Politics of Numbers,* pp. 415–447.

115 *The vaults of the Smithsonian were once open:* Edward Wyatt, "Smithsonian-Showtime TV Deal Raises Concerns, *New York Times,* March 31, 2006.

"optioned America's attic": Edward Wyatt, "Smithsonian Agreement Angers Filmmakers," *New York Times,* April 1, 2006.

selling off public assets: Dennis Cauchon, "States to Sell or Lease Assets," *USA Today,* June 8, 2006.

tax collection . . . by casinos: Dirk Johnson, "Gambling's Spread: Gold Rush or Fool's Gold," *New York Times,* October 6, 1991; Michael deCourcy Hinds, "Riverboat Casinos Seek a Home in Pennsylvania," *New York Times,* April 7, 1994; Leslie Eaton, "Tax Revenues Are a Windfall For Louisiana," *New York Times,* June 26, 2006.

pay the IRS with credit cards: Background information is available from the IRS at www.irs.gov/efile/article/o,,id=101316,00.html.

official interrogations . . . have been outsourced: Doug Struck, "Canadian Was Falsely Accused, Panel Says," *Washington Post,* September 19, 2006; Alan Cowell, "British Role in U.S. Policy on Detainees Raises Storm," *New York Times,* January 20, 2006; Ian Fisher, "Reports of Secret U.S. Prisons in Europe Draw Ire and Otherwise Red Faces," *New York Times,* December 1, 2005.

sale of naming rights: Anne Schwartz, "Naming For Dollars," Gothamgazette.com, May 13, 2002; see also the Web site Namingrights online.com.

116 *the Department of the Interior has proposed:* Mark Clayton, "America's National Parks: No Longer Ad-Free Zones?" *Christian Science Monitor,* March 31, 2006.

"privatization rate": Cullen Murphy, "Feudal Gestures," *Atlantic Monthly,* October 2003.

On paper the federal work force: Light, *True Size of Government,* pp. 1–6.

117 *"Contractors have become so big":* Jane Mayer, "Contract Sport," *The New Yorker,* February 16, 2004.

Caesar had such a person: Whittaker, *Frontiers of the Roman Empire,* pp. 108–110.

Halliburton itself used subsidiaries: Colum Lynch, "Firm's Iraq Deals

Greater Than Cheney Has Said," *Washington Post,* June 23, 2001; Jane Mayer, "Contract Sport," *The New Yorker,* February 16, 2004.

117   *overcharged the government:* Jane Mayer, "Contract Sport," *The New Yorker,* February 16, 2004.

bribes-and-contracts conspiracy: James Glanz, "Wide Plot Seen in Guilty Plea in Iraq Project," *New York Times,* February 2, 2006.

118   *a scathing report:* James Glanz, "Iraq Rebuilding Badly Hobbled, U.S. Report Finds," *New York Times,* January 24, 2006; James Glanz, "Idle Contractors Add Millions to Rebuilding," *New York Times,* October 25, 2006.

"spy drain": Walter Pincus, "Increase in Contracting Intelligence Jobs Raises Concerns," *Washington Post,* March 20, 2006.

at least ninety former top officials: Eric Lipton, "High Contractor Pay Lures Counterterrorism Officials," *New York Times,* June 18, 2006.

"Everyone I knew": Robert Baer, *Blow the House Down,* p. 88.

The company that fits this profile: Donald L. Barlett and James B. Steele, "The $8 Billion Shadow," *Vanity Fair,* February 2007; Shane Harris, "'A Stealth Company,'" *National Journal,* July 8, 2006.

119   *"We used to joke":* Shane Harris, "'A Stealth Company,'" *National Journal,* July 8, 2006.

turning the job of border police over to multinational contractors: Eric Lipton, "Bush Turns to Big Military Contractors for Border Control," *New York Times,* May 18, 2006.

4. THE OUTSIDERS

121   *"When given the German command":* Velleius Paterculus, *Roman History,* quoted in Williams, *Romans and Barbarians,* p 94.

"What struck me most about the palace": Diamond, *Squandered Victory,* p. 298.

122   *"The dismal tract":* Tacitus, *Annals,* 1.61.

"Let us be soon gone": Quoted in Clunn, *Quest for the Lost Roman Legions,* p. xxxviii.

123   *The great physician:* Balsdon, *Romans & Aliens,* p. 60.

a British officer stationed in Germany: Clunn, *Quest for the Lost Roman Legions,* pp. 1–85; Wells, *Battle That Stopped Rome,* pp. 37–55.

124   *"somewhat ponderous":* Velleius Paterculus, *Roman History,* quoted in Williams, *Romans and Barbarians,* p. 94.

that the Germans were manageable: In addition to Clunn, comprehensive accounts of the Teutoburg Forest disaster include Wells, *Battle That Stopped Rome*; Williams, *Romans and Barbarians*, pp. 65–115; and Delbruck, *Barbarian Invasions*, pp. 69–148. See also Fergus M. Bordewich, "The Ambush That Changed History," *Smithsonian*, September 2005.

"maximum ferocity": Velleius Paterculus, *Roman History*, quoted in Williams, *Romans and Barbarians*, p. 94.

125   *"having a hard time of it"*: Cassius Dio, *Roman History*, 56.20.1–2.

"While the Romans were in such difficulties": Ibid., 56.20.4.

a computer game: www.3dgamers.com/games/rome/.

126   *feelings of omnipotence*: Wells, *Battle That Stopped Rome*, p. 216.

inflamed by omens: Cassius Dio, *Roman History*, 56.24.1–5.

Augustus disbanded the German cavalry: Suetonius, *Lives of the Caesars*, "The Deified Augustus," 23, 25, 49.

127   *An enormous monument to Arminius*: Clunn, *Quest for the Lost Roman Legions*, pp. 341–344.

"Underestimation of space": Williams, *Romans and Barbarians*, p. 86.

"Romans simply could not believe": Wells, *Battle That Stopped Rome*, p. 216.

"would not have been receptive": Ibid.

128   *hitting his head against a door*: Suetonius, *Lives of the Caesars*, "The Deified Augustus," 23.

Right after the campy bath scene: Alison Futrell, "Seeing Red: Spartacus as Domestic Economist," in Joshel, Malamud, and McGuire, *Imperial Projections*, pp. 77–118.

"We're an empire now": Ron Suskind, "Faith, Certainty, and the Presidency of George W. Bush," *New York Times Magazine*, October 17, 2004.

symbols . . . like the well-known fasces: Anthony J. Marshall, "Symbols and Showmanship in Roman Public Life: The Fasces," *Phoenix* 38, no. 2 (Summer 1984), pp. 120–141.

129   *"a portable kit"*: Ibid.

"almost never cautious": Goldsworthy, *Roman Army at War*, p. 286.

"the old aggressive culture": Lendon, *Soldiers & Ghosts*, p. 221.

that same Crassus: Plutarch, *Lives*, vol. 3, pp. 315–437.

"desiring for his part": Cassius Dio, *Roman History*, 40.12.1.

130   *"poured molten gold into his mouth"*: Ibid., 40.27.2–3.

conduct a mock parade: Plutarch, *Lives*, vol. 3, p. 399.

the disaster at Cannae: Lendon, *Soldiers & Ghosts*, p. 200.

130 "advanced so far": Polybius, Histories, 3.107–118.
131 "Nor did the Romans entirely learn their lesson": Lendon, Soldiers &
     Ghosts, p. 202.
     The emperor would have none of it: Ammianus Marcellinus, Roman
     History, 31.12.3; Lendon, Soldiers & Ghosts, p. 306.
     As the chronicler Josephus wrote: Quoted in Lendon, Soldiers & Ghosts,
     p. 307.
     "All the Roman commanders": Macdowall, Adrianople, p. 88.
132 The battle was joined: Much about the battle remains confusing. See
     Delbruck, Barbarian Invasions, pp. 269–284; Burns, Barbarians Within
     the Gates, pp. 1–42; Potter, Roman Empire at Bay, pp. 529–532; Ferrill,
     Fall of the Roman Empire, pp. 59–64; Lendon, Soldiers & Ghosts, pp.
     305–309; Macdowall, Adrianople, passim.
     "except the one at Cannae": Ammianus Marcellinus, Roman History,
     p. 31.14.19.
     "Citty upon a Hill": John Winthrop, "A Modell of Christian Charity"
     (1630), in Gunn, ed., Early American Writing, p. 11.
     "thought of themselves not as a single ethnic group": Gary B. Miles,
     "Roman and Modern Imperialism," Comparative Studies in Society
     and History 32, no. 4 (October 1990), pp. 629–659.
     granted Roman citizenship: Potter, Roman Empire at Bay, pp. 138–139.
     the dress of faraway peoples: Brown, World of Late Antiquity, p. 21;
     Balsdon, Romans & Aliens, p. 221.
133 a new picture of "Betty Crocker": Rochelle L. Stanfield, "Blending of
     America," National Journal, September 13, 1997.
     Think schematically: An economic version of the cultural idea is ad-
     vanced in Cunliffe, Greeks, Romans, & Barbarians, p. 3.
     the word "barbarians": Ferris, Enemies of Rome, p. 4.
     One way of translating it: John E. Coleman, "Ancient Greek Ethno-
     centrism," in Coleman and Walz, eds., Greeks and Barbarians, p. 178.
     an anonymous writer in the fourth century: Thompson, Roman Re-
     former, p. 113.
     "May Jesus protect the world": Ferrill, Fall of the Roman Empire, p. 91.
134 "Their decisions were based": Mattern, Rome and the Enemy, p. 70.
     "The Romans acquired information": Hagith S. Sivan, review of Infor-
     mation and Frontiers: Roman Foreign Relations in Antiquity, by A. D.
     Lee, Speculum 71, no. 4 (October 1996), pp. 973–975.
     it's fascinating to watch as the Romans: Heather, Fall of the Roman Em-
     pire, pp. 145–158.

134 *The overland trip:* Fergus Millar, "Emperors, Frontiers, and Foreign Relations, 31 B.C. to A.D. 378," *Britannia* 13 (1982), pp. 1–23.

135 *American intelligence often suffers:* National Commission on Terrorist Attacks, pp. 267–272.

Romans who affected Greek ways: Joseph Ward Swain, "The Theory of the Four Monarchies: Opposition History Under the Roman Empire," *Classical Philology* 35, no. 1 (January 1940), pp. 1–21.

Other groups fared less well: Balsdon, *Romans & Aliens*, pp. 65–67.

The Egyptians: Dorothy J. Thompson, "Egypt and Parthia Through Roman Eyes," *Classical Review* 39, no. 1 (1989), pp. 86–87.

Jews were regarded: Jerry L. Daniel, "Anti-Semitism in the Hellenic-Roman Period," *Journal of Biblical Literature* 98, no. 1 (March 1979), pp. 45–65; Fergus Millar, "Last Year in Jerusalem: Monuments of the Jewish War in Rome," in Edmondson, Mason, and Rives, eds., *Flavius Josephus and Flavian Rome*, pp. 101–128.

136 *"A German is not":* Tacitus, *Germania*, 14.

Cicero was quick to point out: Quoted in Balsdon, *Romans & Aliens*, p. 2.

You can't miss an echo: John Hendren, "General's Speeches Broke Pentagon Rules," *Los Angeles Times*, August 20, 2004; editorial, "A General's Religious War," *Chicago Tribune*, August 28, 2004.

137 *Strabo and Plutarch:* Balsdon, *Romans & Aliens*, pp. 201–206.

"on some of these subject nations": Appian, *Roman History*, Preface, 7.

the world's lingua franca: Barber, *Jihad vs. McWorld*, p. 84.

138 *America has confronted the hostility:* See, for instance. Fareed Zakaria, "The Politics of Rage: Why Do They Hate Us?" *Newsweek*, October 15, 2001; Michael Kamber, "Why They Hate Us," *Village Voice*, October 16, 2001; Jonathan Tepperman, "The Anti-Anti-Americans," *New York Times*, December 12, 2004; Ivan Eland, "It's What We Do," *American Prospect*, January 2006.

Europeans are considerably more likely: "America's Image Slips, but Allies Share U.S. Concerns over Iran, Hamas," Pew Global Attitudes Project (Washington, DC: Pew Research Center, 2006).

"acid reflux": Margaret Drabble, "I Loathe America, and What It Has Done to the Rest of the World," *Daily Telegraph*, August 5, 2003.

Harold Nicolson told a friend: Patten, *Cousins and Strangers*, p. 4.

"vassals and tributaries": Brzezinski, *Grand Chessboard*, p. 6.

139 *Opinion surveys show:* "America's Image Slips, but Allies Share U.S.

Concerns over Iran, Hamas," Pew Global Attitudes Project (Washington, DC: Pew Research Center, 2006).

139 *a columnist in our stalwart ally Britain:* Charlie Brooker, "Dumb Show," *The Guardian,* October 23, 2004.

publicly degrading the captured leaders: Michael Vlahos, "The Weakness of Empire," *American Conservative,* May 22, 2006.

The story is told: The king was Antiochus IV of Syria. Eliot Cohen, "History and the Hyperpower," *Foreign Affairs,* July/August 2004; Mattern, *Rome and the Enemy,* p. 213. See also Balsdon, *Romans & Aliens,* p. 171.

140 *Polybius . . . refers to the Romans:* Craige Champion, "Romans as *Barbaroi:* Three Polybian Speeches and the Politics of Cultural Indeterminacy," *Classical Philology* 95 (October 2000), pp. 425–444.

the annual Teamsters convention: Margaret Malamud and Donald T. MacGuire, Jr., "Living Like Romans in Las Vegas," in Joshel, Malamud, and McGuire, eds., *Imperial Projections,* p. 256.

one wealthy Roman: Balsdon, *Romans & Aliens,* p. 177.

The Greek philosopher Demonax: Ibid., p. 178.

141 *"It was possible for an American":* Quoted in Stanley Meisler, "A Nation of No-Nothings," *Los Angeles Times,* December 2, 1990.

Lynne Cheney wondered aloud: Remarks at the Dallas Institute of Humanities and Culture, October 5, 2001; office of Lynne Cheney, the White House.

cannot name the ocean: Asia in the Schools: Preparing Young Americans for Today's Interconnected World (New York: Asia Society, 2001).

can't locate Iran or Iraq: National Geographic/Roper 2002 Global Geographic Literacy Survey (Washington, DC: National Geographic Education Foundation, 2002).

"stubborn monolingualism": Securing America's Future: Global Education for a Global Age (Washington, DC: NAFSA: Association of International Educators, 2003).

to get from Japan to Australia: National Geographic/Roper Public Affairs, 2006 Geographic Literacy Study (Washington, DC: National Geographic Education Foundation, 2006).

142 *severe restrictions on foreign students:* Burton Bollag, "College Officials Report Frustrations with Homeland-Security Agents," *Chronicle of Higher Education,* June 17, 2005.

applications have slipped sharply: Sam Dillon, "U.S. Slips in Attracting the World's Best Students," *New York Times,* December 21, 2004.

142   *three minutes to the genocide in Darfur*: Sherry Ricchiardi, "Déjà Vu," *American Journalism Review*, February/March 2005.

"I once asked an American general": H.D.S. Greenway, "Heeding British Ghosts," *Boston Globe*, June 6, 2006.

so-called "black budget": Scott Shane, "Official Reveals Budget for U.S. Intelligence," *New York Times*, November 8, 2005.

143   *"Sterling exceptions aside"*: Edward G. Shirley [Reuel Marc Gerecht], "Can't Anybody Here Play This Game?" *Atlantic Monthly*, February 1998.

"moral barrier" . . . "information barrier": A. Alfoldi, "The Moral Barrier on Rhine and Danube," *Congress of Roman Frontier Studies* (Durham, 1952), pp. 1–16; Fergus Millar, "Emperors, Frontiers, and Foreign Relations, 31 B.C. TO A.D. 378," *Britannia* 13 (1982), pp. 1–23.

"an ideology of the foreigner": Mattern, *Rome and the Enemy*, p. 76.

"reality-based community": Ron Suskind, "Faith, Certainty, and the Presidency of George W. Bush," *New York Times Magazine*, October 17, 2004.

the motivations of his main character: Greene, *Quiet American*, p. 13.

144   *two basic components of the American stance*: Hartz, *Liberal Tradition in America*, pp. 285–286.

145   *"We came in peace"*: Godwin, *Apollo 11*, p. 11.

Romans sought "symbolic deference": Mattern, *Rome and the Enemy*, p. 162.

laid down a single criterion: Bremer, *My Year in Iraq*, p. 76.

"needs to clearly and publicly express": Thom Shanker and Mark Mazzetti, "Bush Said to Be Frustrated by Level of Public Support in Iraq," *New York Times*, August 16, 2006.

a fictitious ethnic group: "American Notes: Ethnicity," *Time*, January 20, 1992; American Jewish Committee.

146   *The "just like us" argument*: Benjamin Schwarz, "The Diversity Myth," *Atlantic Monthly*, May 1995.

sameness of popular commerce and culture: Barber, *Jihad vs. McWorld*, pp. 17–18, 83–84.

most of the wealth, creativity, and entrepreneurship: Richard Florida, "The World Is Spiky," *Atlantic Monthly*, October 2005.

Roman Empire . . . urbanized and spiky: Jones, *Later Roman Empire*, pp. 1021–1022.

"a sharp cultural cleavage": Ibid., p. 995

147   *"The empire was ruled"*: Brown, *The World of Late Antiquity*, p. 14.

148 *he was routinely referred to in the press:* See, for instance, Steve and Cokie Roberts, "Political Interests Blind Bush to Military Disaster," *Chicago Sun-Times*, April 11, 2004; Asia Aydintasbas, "Turning Friend into Foe in Baghdad," *New York Times*, May 22, 2004.

Prior to taking up his post: Bremer, *My Year in Iraq*, p. 4.

translators, scarce at the outset: Katherine McIntire Peters, "Lost in Translation," *Government Executive*, May 2002; Hendrik Hertzberg, "Studies Say," *New Yorker*, December 18, 2006; Renee Merle, "First Ears, Then Hearts and Minds," *Washington Post*, November 2, 2006.

"From inside the palace": Diamond, *Squandered Victory*, p. 92.

"Mission Accomplished" action figure: Ibid., p. 75.

"fractal set of hierarchies": Maier, *Among Empires*, pp. 60–61.

to create an embryonic version: Peter Galbraith, "The Mess," *New York Review of Books*, March 9, 2006.

"I hoped that these sessions would evolve": Bremer, *My Year in Iraq*, p. 63.

a traffic code: William Langewiesche, "Welcome to the Green Zone," *Atlantic Monthly*, November 2004. See also Peter Galbraith, "The Mess," *New York Review of Books*, March 9, 2006. Diamond, *Squandered Victory*, provides personal details on American operations inside the Green Zone. The best book-length treatment of the subject is Chandrasekaran, *Imperial Life in the Emerald City*.

One Iraqi employed in the Green Zone: Muean Aljabiry, "How Do You Say Clueless?" *Washington Post*, March 19, 2006.

150 *fiasco in Somalia:* Craig Timberg, "Mistaken Entry into Clan Dispute Led to U.S. Black Eye in Somalia," *Washington Post*, July 2, 2006.

"demonstrations of will": Charles Krauthammer, "The Bush Doctrine," *Time*, March 5, 2001.

151 *trained and armed Muslim warriors:* Mary Anne Weaver, "Blowback," *Atlantic Monthly*, May 1996.

encouraged . . . to start growing flowers: Christopher S. Wren, "U.S. Saves on Flowers from Andes," *New York Times*, February 17, 1997; Anthony DePalma, "In Trade Issue, the Pressure Is on Flowers," *New York Times*, January 24, 2002.

giving rise to . . . Korean words: Yongshik Bong, "Yongmi: Pragmatic Anti-Americanism in South Korea," *Brown Journal of World Affairs* 10, no. 2 (Winter/Spring 2004), pp. 153–165.

area codes: Ariana Eonjung Cha, "Baghdad's U.S. Zone a Stand-in for

Home," *Washington Post*, December 6, 2003; Robin Roberts, "One Year Later," *Good Morning America*, ABC News, March 11, 2004; William Langewiesche, "Welcome to the Green Zone," *Atlantic Monthly*, November 2004.

5. THE BORDERS

152 *"The barbarians were adapting themselves"*: Cassius Dio, *Roman History*, 56.18.2.

"Ai pledch aliyens": Editorial, "People Power," *New York Times*, April 12, 2006.

the great stonework fortification: Breeze and Dobson, *Hadrian's Wall*, pp. 25–43; Burton, *Hadrian's Wall Path*, pp. 16–23.

153 *"The hard road goes on and on"*: Kipling, *Puck of Pook's Hill*, p. 146.

154 *ordnance map reveals*: Ordnance Survey/OL43: Hadrian's Wall, 2005.

155 *did their own heavy lifting*: Breeze and Dobson, *Hadrian's Wall*, pp. 66–79.

nineteenth-century calculation: Ibid., pp. 82–83.

the total population: Goodman, *Roman World*, 44 B.C.–A.D. 180, p. 159.

157 *"A border . . . is where you draw a line"*: Dennis West and Joan M. West, "Borders and Boundaries: An Interview with John Sayles," *Cineaste* 22, no. 3 (Summer 1996).

not the exterior but the interior: Charles S. Maier, "An American Empire?" *Harvard*, November/December 2002.

158 *DNA . . . could someday mark a political border*: Peter Prengaman, "DNA Testing More Common Among Immigrants," Associated Press, July 27, 2006.

159 *experts have been gathering regularly*: The proceedings of the *Limes* Congresses are published in *British Archaeological Reports*, Oxford (www.archaeopress.com).

"in a fit of absence of mind": Seeley, *Expansion of England*, quoted in Owen Chadwick, "Historian of Empire," *Modern Asian Studies* 15, no. 4 (1981), pp. 877–880.

Tiberius once urged his stepson: Tacitus, *Annals*, 2.26.

160 *Trajan . . . built an arched span*: Cassius Dio, *Roman History*, 68.13.1–2.

the earthen Antonine Wall: Breeze and Dobson, *Hadrian's Wall*, pp. 88–116.

160 *for one fifty-mile stretch:* Olwen Brogan, "An Introduction to the Roman Land Frontier in Germany," *Greece & Rome* 3, no. 7 (October 1933), pp. 22–30; Peter S. Wells, "The *Limes* and Hadrian's Wall," *Expedition* 47, no. 1, pp. 18–24.

walls and trenches in the middle of nowhere: Williams, *Reach of Rome*, pp. 115–155.

transport planes . . . fly into Rasheed: GlobalSecurity.org.

161 *"All along the borders":* Cambridge Ancient History 11, p. 82.

another authority referred to the line: Andreas Alfoldi, "The Moral Barrier on Rhine and Danube," in Birley, ed., *Congress*, pp. 1–16.

"There was no Roman historian": Ibid.

Domitian, after wiping out one tribe: Mattern, *Rome and the Enemy*, p. 135.

enlisted prominent architects: Linda Hales, "At the Borders, Creative Crossings," *Washington Post*, July 29, 2006.

162 *territorial flux has been the norm:* Whittaker, *Rome and Its Frontiers*, pp. 1–2; International Boundary Research Unit, University of Durham (www.dur.ac.uk/ibru/).

Why did the frontiers stop where they did?: John Cecil Mann, "The Frontiers of the Principate," *Aufstieg und Niedergang der romischen Welt* 2, no. 1 (1974).

Augustus wrote out instructions: Tacitus, *Annals*, 1.11; Mattern, *Rome and the Enemy*, pp. 90–91.

"the view of a weary . . . old man": John Cecil Mann, "The Frontiers of the Principate," *Aufstieg und Niedergang der romischen Welt* 2, no. 1 (1974).

the warning recalls the farewell addresses: Commager, *Documents of American History*, pp. 174, 653.

163 *none that announced the edge of the imperium itself:* Isaac, *Limits of Empire*, pp. 395–398.

borders were established . . . for different reasons: Whittaker, *Frontiers of the Roman Empire*, pp. 8–9, 85–97.

Plenty of emperors after Augustus: Ibid., p. 8.

164 *"shall we go on conferring our Civilization?":* Mark Twain, "To the Person Sitting in Darkness," in Zwick, *Mark Twain's Weapons of Satire*, pp. 22–39.

sales of salsa: Michael J. Weiss, "The Salsa Sectors," *Atlantic Monthly*, May 1997.

you really would see something: MacMullen, *Romanization*, pp. 124–137; Heather, *Fall of the Roman Empire*, pp. 44–45.

165   *something akin to dog tags:* Davies, *Service in the Roman Army,* p. 164.
some 20,000 identical statues: MacMullen, *Romanization,* p. 129.
after a coup dislodged one barbarian: Tacitus, *Annals,* 2.62.
stout, swift horses . . . frozen remains: John W. Eadie, "The Development of Roman Mailed Cavalry," *Journal of Roman Studies* 57, no. 1/2 (1967), pp. 161–173.
Map the places where Roman artifacts have been found: Whittaker, *Frontiers of the Roman Empire,* pp. 98–131; Cunliffe, *Greeks, Romans & Barbarians,* pp. 171–192.

166   *German gets its word:* Williams, *Romans and Barbarians,* pp. 75, 77.
only word left by the Visigoths: Brown, *World of Late Antiquity,* p. 125.
barbarians were fast learners: Whittaker, *Frontiers of the Roman Empire,* p. 118.
Attila employed several Latin secretaries: Thompson, *Huns,* pp. 139–140.
The one eyewitness account: Ibid., pp. 122–123.
clamp down on trade: Whittaker, *Frontiers of the Roman Empire,* p. 119; Elton, *Frontiers of the Roman Empire,* p. 89.
channel commerce . . . weren't passports: Whittaker, *Rome and Its Frontiers,* pp. 204–205; Whittaker, *Frontiers of the Roman Empire,* p. 121.

167   *"undramatic adjustments":* Goffart, *Barbarians and Romans,* p. 4.
somewhat Romanized barbarians: Whittaker, *Rome and Its Frontiers,* pp. 130, 204–208.
It has been said of Kipling: Whittaker, *Frontiers of the Roman Empire,* p. 2.

168   *his family . . . exemplifies how quickly:* Elton, *Frontiers of the Roman Empire,* p. 38; Whittaker, *Rome and Its Frontiers,* p. 209.
Think of Stilicho . . . Alaric himself had once served Rome: Whittaker, *Rome and Its Frontiers,* pp. 212–213; Burns, *Barbarians Within the Gates,* pp. 183–223.
comes from maps in textbooks . . . a sense of unstoppable power: Walter Goffart, "Rome, Constantinople, and the Barbarians," *American Historical Review* 86, no. 2 (April 1981), pp. 275–306.

169   *a long reflection about the 1992 Los Angeles riots:* Jack Miles, "Blacks vs. Browns," *Atlantic Monthly,* October 1992.
usually with time to recover . . . Some cities: Whittaker, *Rome and Its Frontiers,* pp. 50–57.
not above picking fights: John F. Drinkwater, "'The Germanic Threat on the Rhine Frontier': A Romano-Gallic Artefact?" in Mathisen and Sivan, eds., *Shifting Frontiers in Late Antiquity,* pp. 20–30.

169 *"the chance to show their mettle"*: Linda Feldmann, "Presidencies Hewn by War," *Christian Science Monitor*, March 21, 2003.

170 *"carelessly transmitted numerals"*: Whittaker, *Rome and Its Frontiers*, p. 53.

cautions against the traditional rhetoric: Ibid., p. 54. Goffart, in *Barbarians and Romans*, also resists the "vocabulary of floods, waves, and other vivid images" (p. 4).

"probably . . . wrong to estimate": Walter Goffart, "Rome, Constantinople, and the Barbarians," *American Historical Review* 86, no. 2 (April 1981), pp. 275–306.

the actual fighting force in each instance: Whittaker, *Frontiers of the Roman Empire*, p. 212.

invasions should be seen . . . as individual events: Walter Goffart, "Rome, Constantinople, and the Barbarians," *American Historical Review* 86, no. 2 (April 1981), pp. 275–306.

fragmented and fractious: Whittaker, *Frontiers of the Roman Empire*, pp. 212–213.

If there was a tipping point: Peter Heather, "The Huns and the End of the Roman Empire in Western Europe," *English Historical Review* 110, no. 435 (February 1995), pp. 4–41; Walter Goffart, "Rome, Constantinople, and the Barbarians," *American Historical Review* 86, no. 2 (April 1981), pp. 275–306.

171 *"no major defeat was suffered by the Roman army"*: Whittaker, *Rome and Its Frontiers*, p. 52; Peter Heather, "The Huns and the End of the Roman Empire in Western Europe," *English Historical Review* 110, no. 435 (February 1995), pp. 4–41.

"concessions to barbarians were safer": Goffart, *Barbarians and Romans*, p. 34.

It meant giving up the revenues: Peter Heather, "The Huns and the End of the Roman Empire in Western Europe," *English Historical Review* 110, no. 435 (February 1995), pp. 4–41.

"What we call the Fall of the Western Roman Empire": Goffart, *Barbarians and Romans*, p. 35.

172 *environmentalists lament that the river*: Richard Bernstein, "No Longer Europe's Sewer, but Not the Rhine of Yore," *New York Times*, April 21, 2006.

174 *"There was a strong feeling that federal agencies"*: William Finnegan, "The Terrorism Beat," *The New Yorker*, July 25, 2005.

175 *Chinese ownership of an important American energy company*: Alexei

Barrionuevo, "Foreign Suitors Nothing New in U.S. Oil Patch," *New York Times*, July 1, 2005.

176 *not hard to envision how a modern Asian catalyst:* One plausible scenario is described in James Fallows, "Countdown to a Meltdown," *Atlantic Monthly*, July–August 2005.

the biography of a counterfeit Prada handbag: Moises Naim, "It's Not About Maps," *Washington Post* ("Outlook"), May 28, 2006.

177 *our Rhine and Danube frontier . . . running nearly 2,000 miles:* Angie C. Marek, "Border Wars," *U.S. News & World Report*, November 28, 2005.

178 *"the largest between any two contiguous countries":* Quoted in Fareed Zakaria, "To Become an American," *Newsweek*, April 10, 2006.

"potentially explosive . . . revolution like no other": Hanson, *Mexifornia*, pp. 17, 142.

Immigrants are choosing new destinations: William H. Frey, *Diversity Spreads Out: Metropolitan Shifts in Hispanic, Asian, and Black Populations Since 2000* (Washington, DC: Brookings Institution, 2006); Rick Lyman, "New U.S. Immigrants Fan Out Across Nation," *New York Times*, August 15, 2006.

179 *"The Roman empire of the fourth century":* Whittaker, *Rome and Its Frontiers*, p. 203.

"Let the prisoners pick the fruits": Quoted in David Brooks, "Scuttling Toward Sanity," *New York Times*, April 6, 2006.

Louisiana's "rent a convict" practice: Adam Nossiter, "With Jobs to Do, Louisiana Parish Turns to Inmates," *New York Times*, July 5, 2006.

a Spanish-language recording: David Montgomery, "An Anthem's Discordant Notes," *Washington Post*, April 28, 2006.

"The images that sprang into my mind": Roger Hernandez, "Anthem Top Topic So Far for Readers," *El Paso Times*, August 22, 2006.

180 *capable of supplying . . . bean-and-beef burritos:* P. W. Singer, "A Run for the Border," *Washington Post* ("Outlook"), July 9, 2006.

built and operated by private corporate security forces: Spencer Hsu and John Pomfret, "Technology Has Uneven Record on Securing Border," *Washington Post*, May 21, 2006.

barbarians being slaughtered . . . barbarians being welcomed: Whittaker, *Rome and Its Frontiers*, p. 204.

campaign film . . . waves a small Mexican flag: Richard Rodriguez, "What a Wall Can't Stop," *Washington Post*, May 28, 2006.

181   *"We've done open arms"*: Siobhan Gorman, "Immigration: The End-less Flood," *National Journal*, February 7, 2004.

"millions of barbarians had been pacified": Walter Goffart, "Rome, Constantinople, and the Barbarians," *American Historical Review* 86, no. 2 (April 1981), pp. 275–306. See also Whittaker, *Rome and Its Frontiers*, p. 202.

182   *half . . . would die before achieving that goal*: Mattern, *Rome and the Enemy*, p. 86.

accommodate the waves of immigration: Yearbook of Immigration Statistics, Office of Immigration Statistics, United States Department of Homeland Security, September 2003.

In the earliest days of Ellis Island: Michael Powell, "U.S. Immigration Debate Is a Road Well Traveled," *Washington Post*, May 8, 2006.

radio stations are going bilingual . . . Hispanic high school graduates prefer English: Richard Cohen, "My History of English-Only," *Washington Post*, May 30, 2006; Joel Kotkin, "The Multiculturalism of the Streets," *American Interest*, Spring 2006; Martin Miller, "It's Pure Spanglish at This L.A. Radio Station," *Los Angeles Times*, November 5, 2005.

183   *"If you ask a second-generation American Muslim"*: Quoted in James Fallows, "Declaring Victory," *Atlantic Monthly*, September 2006.

rates of intermarriage: Joel Kotkin, "The Multiculturalism of the Streets," *American Interest*, Spring 2006.

Population projections suggest: James P. Smith and Barry Edmonston, "The New Americans: Economic, Demographic, and Fiscal Effects of Immigration" (Washington, DC: National Research Council of the National Academy of Sciences, 1997).

"how it will affect . . . Miss America": Siobhan Gorman, "Immigration: The Endless Flood," *National Journal*, February 7, 2004.

Another analyst, who studies the "great immigrant portals": Joel Kotkin, "The Multiculturalism of the Streets," *American Interest*, Spring 2006.

thousands . . . united to say these words: Editorial, "People Power," *New York Times*, April 12, 2006.

184   *a masterly and influential study*: The reference is to Brown's *World of Late Antiquity*.

left its mark on him: Peter L. Brown, "The World of Late Antiquity Revisited," *Symbolae Osloenses* 72 (1997), pp. 5–30.

"without invoking an intervening catastrophe": Peter L. Brown, "The

World of Late Antiquity Revisited," *Symbolae Osloenses* 72 (1997), pp. 5–30.

EPILOGUE: THERE ONCE WAS A GREAT CITY

185 *"One thought alone"*: Coetzee, *Waiting for the Barbarians*, p. 131.
capital . . . moved there from Rome: Ferrill, *Fall of the Roman Empire*, p. 99. Also Lewis, *Dante*, pp. 190–191.

187 *"extreme recalcitrance of the evidence"*: Ramsay MacMullen, "Roman Elite Motivation: Three Questions," *Past and Present* 88 (1980), pp. 3–16.
Nero, famous for playing the cithara: C. E. Manning, "Acting and Nero's Conception of the Principate," *Greece & Rome* 22, no. 2 (October 1975), pp. 164–175; Holland, *Nero*, p. 70; Goodman, *Roman World*, pp. 4–9.
What we know about Romulus Augustulus is this: Geoffrey Nathan, "The Last Emperor: The Fate of Romulus Augustulus," *Classica et Mediaevalia* 43 (1992), pp. 261–271; Ralph W. Mathisen and Geoffrey Nathan, "Romulus Augustulus (475–476 A.D.) — Two Views," *De Imperatoribus Romanis: An Online Encyclopedia of Roman Emperors*, www.roman-emperors.org/. See also Bury, *Invasion of Europe*, pp. 166–183; Gibbon, *Decline and Fall*, vol. 2, pp. 342–344.

188 *seems to have left Ravenna . . . given a pension*: Geoffrey Nathan, "The Last Emperor: The Fate of Romulus Augustulus," *Classica et Mediaevalia* 43 (1992), pp. 261–271.
The most obvious route: T. Ashby and R.A.L. Fell, "The Via Flaminia," *Journal of Roman Studies* 11 (1921), pp. 125–190; Chevalier, *Roman Roads*, pp. 131–139.
That whole stretch of the Italian shore: For a memorable evocation of this region, see Robert Harris's historical novel *Pompeii*. For resort life specifically, see Perrottet, *Pagan Holiday*, pp. 63–90.
the truly rich were known: Casson, *Travel in the Ancient World*, p. 141.
Hadrian died somewhere on this hillside: Danziger and Purcell, *Hadrian's Empire*, p. 283.
has described the estate of Lucullus: Plutarch, *Lives*, vol. 2, pp. 602–603.
Romulus Augustulus apparently passed his days: Geoffrey Nathan, "The Last Emperor: The Fate of Romulus Augustulus," *Classica et Mediaevalia* 43 (1992), pp. 261–271.

189 *Odoacer was not so fortunate:* Gibbon, *Decline and Fall,* vol. 2, pp. 452–453.

Others made the sensible point: Richard Parker, "Inside the 'Collapsing' Soviet Economy," *Atlantic Monthly,* June 1990.

190 *"no proof that any important skills . . . were lost":* Lynn White Jr., "Technology and Invention in the Middle Ages," *Speculum* 15, no. 2 (April 1940), pp. 141–159.

191 *"After the initial shock of barbarian incursions":* Gary B. Miles, "Roman and Modern Imperialism: A Reassessment," *Comparative Studies in Society and History* 32, no. 4, (October 1990), pp. 629–659.

A senate composed of aristocrats: Lancon, Rome in Late Antiquity, pp. 48–53.

a general subsidence in well-being over time: Ward-Perkins, *Fall of Rome,* pp. 104–120.

"empire without end": Virgil, *Aeneid,* 1.279.

Virgil was speaking of Troy: Ibid., 2.363.

"translation of empire": Nordholt, *Myth of the West,* pp. 2, 6, 190–207.

192 *"The world's scepter passed":* Ibid., p. 202.

Charles Darwin joined in: Ibid., p. 196.

"There is the moral of all human tales": Byron, *Childe Harold's Pilgrimage,* in *Major Works,* p. 179.

193 *its official two-tier system:* Jones, *Later Roman Empire,* pp. 516–522; Gary B. Miles, "Roman and Modern Imperialism," *Comparative Studies in Society and History* 32, no. 4 (October 1990), pp. 629–659; Ramsay MacMullen, "Personal Power in the Roman Empire," *American Journal of Philology* 107, no. 4 (Winter 1986), pp. 512–524.

"Nothing is more unfair than equality": Quoted in Robert J. Antonio, "The Contradiction of Domination and Production in Bureaucracy: The Contribution of Organizational Efficiency to the Decline of the Roman Empire," *American Sociological Review* 44, no. 6 (December 1979), pp. 895–912.

194 *the rich got their way in the manner of warlords:* MacMullen, *Roman Social Relations,* pp. 8, 42, 112.

Rome ran on slaves: W. V. Harris, "On War and Greed in the Second Century B.C.," *American Historical Review* 76, no. 5 (December 1971), pp. 1371–1385; Schiavone, *End of the Past,* pp. 120–122.

may have made up half the total population: Mattern, *Rome and the Enemy,* p. 153.

194 *a prominent politician . . . killed by one of his slaves:* Pliny, *Natural History*, 10.16; Tacitus, *Annals*, 14.40.

195 *"It is true that, because of this":* Schiavone, *End of the Past*, p. 111.

"the tranquil and prosperous state of the empire": Gibbon, *Decline and Fall*, vol. 1, p. 50.

196 *imagines the journey his poem will take:* Ovid, *Tristia*, 1.1.1–2.

on the second floor: Suetonius, *Lives of the Caesars,* "The Deified Augustus," 72; Nunzio Giustozzi, "Apollo in the House of Augustus: Living with the God," in La Regina, ed., *Archaeological Guide to Rome*, pp. 58–59.

The Domus Flavia sprawls nearby: La Regina, ed., *Archaeological Guide to Rome*, pp. 48–76.

198 *made a start on each one:* Some of what follows grew out of conversations over more than a decade with my colleague Robert D. Kaplan.

199 *"If U.S. metropolitan areas were countries":* Richard Florida, "The World Is Spiky," *Atlantic Monthly*, October 2005.

200 *of the world's hundred largest "economies":* Robert D. Kaplan, "Was Democracy Just a Moment?" *Atlantic Monthly*, December 1997.

201 *"Herein lies one of the curses of empire":* Eliot Cohen, "History and the Hyperpower," *Foreign Affairs*, July 2004.

Nero was castigated: Everett L. Wheeler, "Methodological Limits and the Mirage of Roman Strategy, Part I," *Journal of Military History* 57, no. 1 (January 1993), pp. 7–41.

empires . . . decline when they cease to expand: Thucydides, *History of the Peloponnesian War*, 6.18.

a lot of academic theorizing: For a brief survey, see Emily S. Rosenberg, "Bursting America's Imperial Bubble," *Chronicle Review*, November 3, 2006.

"Species, people, firms, governments": Ormerod, *Why Most Things Fail*, p. 221.

202 *"We may say that Symmachus and his friends":* Barrow, *Prefect and Emperor*, p. 14.

In the realm of foreign policy alone this could mean: Robert Kagan, quoted in Charles William Maynes, "The Perils of (and for) an Imperial America," *Foreign Policy*, no. 111 (Summer 1998), pp. 36–48.

the more powerful America becomes: Niebuhr, *Irony of American History*, p. 74.

203 *"An empire remains powerful":* Quoted in Anthony Pagden, "The Problems of Superpower," *Los Angeles Times*, November 14, 2004.

204 *"Empire-builders yearn for stability"*: Charles S. Maier, "An American Empire?" *Harvard*, November/December 2002.

program of national service: Many proposals have been put forward over the years. See, for instance, Charles Moskos and Paul Glastris, "Now Do You Believe We Need a Draft?" *Washington Monthly*, November 2001.

205 *"ruthlessly extracted"*: Luttwak, *Grand Strategy*, p. 130.

Rome's elites were deeply satisfied: Schiavone, *End of the Past*, pp. 3–15, 175–203.

206 "We have long since": Sallust, *Conspiracy of Cataline*, 52.11.

Bibliography

Abbott, Carl. *Political Terrain: Washington, D.C., from Tidewater Town to Global Metropolis.* Chapel Hill: University of North Carolina Press, 1999.

Adam, J. P. *Roman Building Materials and Techniques.* New York: Routledge, 1994.

Addison, Joseph. *Cato: A Tragedy, and Selected Essays,* edited by Christine Dunn Henderson and Mark E. Yellin. Indianapolis: Liberty Fund, 2004.

Adkins, Leslie, and Roy A. Adkins. *Handbook to Life in Ancient Rome.* New York and Oxford: Oxford University Press, 1994.

Adler, Bill, ed. *Washington: A Reader.* New York: Meredith Press, 1967.

Alonso, William, and Paul Starr, eds. *The Politics of Numbers.* New York: Russell Sage Foundation, 1987.

Ammianus Marcellinus. *Ammianus Marcellinus in Three Volumes, Revised Edition.* Translated by John C. Rolfe. Loeb Classical Library. Cambridge, MA: Harvard University Press, 1952.

Appian. *Roman History.* Translated by Horace White. Loeb Classical Library. Cambridge, MA: Harvard University Press, 1912.

Aristides. *P. Aelius Aristides: The Complete Works.* Vol. 2. Translated by Charles A. Behr. Leiden, Netherlands: E. J. Brill, 1981.

Arnheim, M.T.W. *The Senatorial Aristocracy in the Later Roman Empire.* Oxford: Clarendon Press, 1972.

Auden, W. H. *Collected Poems.* New York: Vintage, 1976.

Augustine. *The City of God.* Introduction by Thomas Merton. New York: Random House, 1950.

Avant, Deborah D. *The Market for Force.* New York and Cambridge: Cambridge University Press, 2005.

Babcock, Michael A. *The Night Attila Died: Solving the Murder of Attila the Hun.* New York: Penguin, 2005.

Baer, Robert. *Blow the House Down.* New York: Crown, 2006.

Balsdon, J.P.V.D. *Romans & Aliens.* London: Duckworth, 1979.

Barber, Benjamin. *Jihad vs. McWorld: Terrorism's Challenge to Democracy.* New York: Random House, 1995.

Barrow, R. H. *Prefect and Emperor.* Oxford: Clarendon Press, 1973.

Behr, Charles A. *Aristides in Four Volumes.* Cambridge: Cambridge University Press, 1972.

Bidwell, P. *Roman Forts in Britain.* London: Batsford/English Heritage, 1997.

Birley, Anthony. *Garrison Life at Vindolanda.* Stroud, Gloucestershire: Tempus, 2002.

Birley, Eric, ed. *The Congress of Roman Frontier Studies, 1949.* Durham: University of Durham, 1952.

Birley, Robin. *The Making of Modern Vindolanda.* Carvoran, Northumberland: Roman Army Museum Publications, 1995.

———. *The Roman Documents from Vindolanda Northumberland.* Carvoran, Northumberland: Roman Army Museum Publications, 1990.

Bowden, Mark. *Guests of the Ayatollah.* New York: Atlantic Monthly Press, 2006.

Bowman, Alan K. *Life and Letters on the Roman Frontier: Vindolanda and Its People.* New York: Routledge, 1994.

Breeze, David J., and Brian Dobson. *Hadrian's Wall.* New York: Penguin, 2000.

Bremer, L. Paul, III. *My Year in Iraq: The Struggle to Build a Future of Hope.* New York: Simon & Schuster, 2006.

Brookhiser, Richard. *Founding Father: Rediscovering George Washington.* New York: Free Press, 1996.

Brown, Peter L. *The World of Late Antiquity.* New York: W. W. Norton, 1989.

Brzezinski, Zbigniew. *The Grand Chessboard.* New York: Basic Books, 1997.

Burns, Thomas S. *Barbarians Within the Gates of Rome.* Bloomington and Indianapolis: Indiana University Press, 1994.

Burton, Anthony. *Hadrian's Wall Path.* London: Aurum, 2004.

Bury, J. B. *The Invasion of Europe by the Barbarians.* New York: W. W. Norton, 1967.

Byron, Lord George Gordon. *Lord Byron: The Major Works.* New York: Oxford University Press, 2000.

Caesar. *The Gallic War.* Translated by H. J. Edwards. Loeb Classical Library. Cambridge, MA: Harvard University Press, 1917.

Carcopino, Jerome. *Daily Life in Ancient Rome.* Edited and annotated by Henry T. Rowell. New Haven, CT: Yale University Press, 2003.

Cassius Dio. *Dio's Roman History.* Vols. 3, 5–8. Translated by Earnest Cary. Loeb Classical Library. Cambridge, MA: Harvard University Press, 1914–1925.

Casson, Lionel. *Travel in the Ancient World.* Toronto: Hakkert, 1974.

Champlin, Edward. *Fronto and Antonine Rome.* Cambridge, MA: Harvard University Press, 1980.

Chandrasekaran, Rajiv. *Imperial Life in the Emerald City: Inside Iraq's Green Zone.* New York: Random House, 2006.

Chevalier, Raymond. *Roman Roads.* Berkeley and Los Angeles: University of California Press, 1976.

Cicero, Marcus Tullius. *Cicero.* Vol. 16. Translated by Clinton Walker Keyes. Loeb Classical Library. Cambridge, MA: Harvard University Press, 1928.

Clunn, Tony. *The Quest for the Lost Roman Legions.* New York: Savas Beatie, 2005.

Coetzee, J. M. *Waiting for the Barbarians.* New York: Penguin, 1980.

Coleman, John E., and Clark A. Walz, eds. *Greeks and Barbarians.* Bethesda, MD: CDL Press, 1997.

Commager, Henry Steele, ed. *Documents of American History.* New York: Appleton-Century-Crofts, 1971.

Cosgrove, Denis, and Stephen Daniels, eds. *The Iconography of Landscape.* New York and Cambridge: Cambridge University Press, 1988.

Coulston, J. C., ed. *Military Equipment and the Identity of Roman Soldiers: Proceedings of the Fourth Roman Military Equipment Conference.* Oxford: British Archaeological Reports, 1988.

Craddock, Patricia B. *Edward Gibbon, Luminous Historian: 1772–1794.* Baltimore: Johns Hopkins University Press, 1989.

Cunliffe, Barry. *Greeks, Romans & Barbarians: Spheres of Interaction.* New York: Methuen, 1988.

Daley, Gregory. *Cannae: The Experience of Battle in the Second Punic War.* New York and London: Routledge, 2002.

Dallek, Robert. *Lyndon B. Johnson, Portrait of a President.* New York: Oxford University Press, 2004.

Damon, Cynthia. *The Mask of the Parasite: A Pathology of Roman Patronage.* Ann Arbor: University of Michigan Press, 1997.

Danziger, Danny, and Nicholas Purcell. *Hadrian's Empire: When Rome Ruled the World.* London: Hodder and Stoughton, 2005.

Davies, Roy W. *Service in the Roman Army*. New York: Columbia University Press, 1989.

Debray, Régis. *Empire 2.0*. Berkeley: North Atlantic, 2004.

Delbruck, Hans. *The Barbarian Invasions*. Translated by Walter J. Renfroe Jr. Lincoln: University of Nebraska Press, 1990.

Diamond, Larry. *Squandered Victory: The American Occupation and the Bungled Effort to Bring Democracy to Iraq*. New York: Times Books, 2005.

Dilke, O.A.W. *Greek & Roman Maps*. Baltimore: Johns Hopkins University Press, 1998.

Dvornik, Francis. *Origins of Intelligence Services*. New Brunswick, NJ: Rutgers University Press, 1974.

Edmondson, Jonathan, Steve Mason, and James Rives, eds. *Flavius Josephus and Flavian Rome*. New York and Oxford: Oxford University Press, 2005.

Elton, Hugh. *Frontiers of the Roman Empire*. Bloomington and Indianapolis: Indiana University Press, 1996.

Epictetus. *The Discourses as Reported by Arrian, The Manual, and Fragments*. Vol. 2. Translated by W. A. Oldfather. Loeb Classical Library. New York: G. P. Putnam's Sons, 1978.

Erdkamp, Paul. *The Roman Army and the Economy*. Amsterdam: Gieben, 2002.

Favro, Diane. *The Urban Image of Augustan Rome*. New York: Cambridge University Press, 1996.

Fehrenbach, T. R. *This Kind of War*. Washington, DC: Brassey's, 1963.

Ferguson, Niall. *Colossus: The Rise and Fall of the American Empire*. New York: Penguin, 2005.

Ferrill, Arther. *The Fall of the Roman Empire*. London: Thames and Hudson, 1986.

———. *Roman Imperial Grand Strategy*. Lanham: University Press of America, 1991.

Ferris, I. M. *Enemies of Rome*. Stroud, Gloucestershire: Sutton, 2000.

Florida, Richard. *The Rise of the Creative Class*. New York: Basic Books, 2002.

Foner, Nancy. *From Ellis Island to JFK: New York's Two Great Waves of Immigration*. New Haven, CT: Yale University Press, 2000.

Freud, Sigmund. *Civilization and Its Discontents*. Introduction by Louis Menand. New York: W. W. Norton, 2005.

Fronto, Marcus Cornelius. *The Correspondence of Marcus Cornelius Fronto*. Translated by C. R. Haines. Loeb Classical Library. New York: G. P. Putnam's Sons, 1970.

Fukuyama, Francis. *The End of History and the Last Man.* New York: Free Press, 1992.

Ganshof, F. L. *Feudalism.* London: Longmans, Green, 1952.

Garreau, Joel. *The Nine Nations of North America.* Boston: Houghton Mifflin, 1981.

Gibbon, Edward. *The Decline and Fall of the Roman Empire.* 3 vols. New York: Modern Library, 1932.

―――. *Memoirs.* Edited by Georges A. Bonnard. New York: Funk & Wagnalls, 1966.

―――. *Private Letters of Edward Gibbon (1753–1794).* London: John Murray, 1896.

Godwin, Robert, ed. *Apollo 11: The NASA Mission Reports.* Vol. 1. Burlington, ON: Apogee Books, 1999.

Goffart, Walter. *Barbarians and Romans, A.D. 418–584: The Techniques of Accommodation.* Princeton, NJ: Princeton University Press, 1980.

Goldsworthy, Adrian Keith. *The Roman Army at War, 100 B.C.–A.D. 200.* Oxford: Clarendon, 1996.

Goodman, Martin. *The Roman World: 44 BC–AD 180.* New York: Routledge, 1997.

Gordon, C. D. *The Age of Attila: Fifth-Century Byzantium and the Barbarians.* Ann Arbor: University of Michigan Press, 1960.

Greene, Graham. *The Quiet American.* New York: Viking, 1955.

Greenfield, Meg. *Washington.* New York: Public Affairs, 2001.

Gross, Robert A. *The Minutemen and Their World.* New York: Hill and Wang, 1976.

Gummere, Richard M. *The American Colonial Mind and the Classical Tradition.* Cambridge, MA: Harvard University Press, 1963.

Gunn, Giles, ed. *Early American Writing.* London: Penguin, 1994.

Guttman, Daniel, and Barry Willner. *The Shadow Government.* With an introduction by Ralph Nader. New York: Pantheon, 1976.

Habinek, Thomas, and Alessandro Schiesaro, eds. *The Roman Cultural Revolution.* New York: Cambridge University Press, 1998.

Hanson, Victor Davis. *Mexifornia: A State of Becoming.* San Francisco: Encounter Books, 2003.

Hardt, Michael, and Antonio Negri. *Empire.* Cambridge, MA: Harvard University Press, 2000.

Hare, Augustus J. C. *Walks in Rome.* New York: Routledge, 1871.

Harris, Robert. *Pompeii.* New York: Random House, 2003.

Hartz, Louis. *The Liberal Tradition in America.* New York: Harcourt, 1955.

Hassall, M.W.C., and Robert Ireland, eds. *De Rebus Bellicis*. Oxford: British Archaeological Reports, 1979.

Heather, Peter. *The Fall of the Roman Empire*. London: Macmillan, 2005.

Heller, Joseph. *Catch-22*. New York: Scribner, 1996.

Hingley, Richard, ed. *Images of Rome: Perceptions of Ancient Rome in Europe and the United States in the Modern Age*. Portsmouth, RI: Journal of Roman Archaeology, 2001.

Holbrooke, Richard. *To End a War*. New York: Random House, 1998.

Holland, Richard. *Nero: The Man Behind the Myth*. London: Sutton, 2000.

Holland, Tom. *Rubicon: The Last Years of the Roman Republic*. New York: Doubleday, 2003.

Holmes, Richard. *Acts of War: The Behavior of Men in Battle*. New York: Macmillan, 1985.

Horace. *The Complete Odes and Epodes*. Translated by W. G. Shepard. London: Penguin, 1983.

Horsley, Richard A. *Jesus and Empire*. Minneapolis: Fortress, 2003.

Hughes, J. Donald. *Pan's Travail: Environmental Problems of the Greeks and Romans*. Baltimore: Johns Hopkins University Press, 1994.

Huntington, Samuel P. *The Soldier and the State: The Theory and Politics of Civil-Military Relations*. Cambridge, MA: Harvard University Press, 1957.

Irons, Peter. *War Powers: How the Imperial Presidency Hijacked the Constitution*. New York: Metropolitan Books, 2005.

Isaac, Benjamin. *The Limits of Empire: The Roman Army in the East*. New York: Oxford University Press, 1990.

Jacob, Kathryn Allamong. *Capital Elites: High Society in Washington, D.C., After the Civil War*. Washington, DC: Smithsonian Institution Press, 1995.

Jacobs, Jane. *Dark Age Ahead*. New York: Random House, 2004.

James, Harold. *The Roman Predicament*. Princeton, NJ: Princeton University Press, 2006.

James, Henry. *The Portrait of a Lady*. New York: Everyman's Library, 1991.

Johnson, Chalmers. *The Sorrows of Empire: Militarism, Secrecy, and the End of the Republic*. New York: Metropolitan Books, 2004.

Jones, A.H.M. *The Later Roman Empire, 284–602*. 2 vols. Baltimore: Johns Hopkins University Press, 1986.

Jones, G.D.B., and D. J. Wooliscroft. *Hadrian's Wall from the Air*. Stroud, Gloucestershire: Tempus, 2001.

Joshel, Sandra, Margaret Malamud, and Donald T. McGuire, Jr., eds. *Imperial Projections*. Baltimore: Johns Hopkins University Press, 2001.

Juvenal. *The Satires of Juvenal.* Translated by Rolfe Humphries. Bloomington: Indiana University Press, 1958.

Kagan, Donald. *The End of the Roman Empire: Decline or Transformation?* Lexington, MA: D. C. Heath, 1978.

Kaplan, Robert D. *Imperial Grunts: The American Military on the Ground.* New York: Random House, 2005.

Kelly, Christopher. *Ruling the Later Roman Empire.* Cambridge, MA: Harvard University Press, 2004.

Kennedy, David, and Derrick Riley. *Rome's Desert Frontier from the Air.* London: B. T. Batsford, 1990.

Kennedy, Paul. *The Rise and Fall of the Great Powers.* New York: Random House, 1987.

Kennon, Donald R., ed. *A Republic for the Ages: The United States Capitol and the Political Culture of the Early Republic.* Charlottesville: University of Virginia Press, 1999.

Keppie, Lawrence. *The Making of the Roman Army: From Republic to Empire.* Norman: University of Oklahoma Press, 1984.

Kipling, Rudyard. *Puck of Pook's Hill.* New York: Penguin, 1988.

Korb, Lawrence J. *Reshaping America's Military.* New York: Council on Foreign Relations, 2003.

Lancon, Bertrand. *Rome in Late Antiquity.* New York: Routledge, 2000.

La Regina, Adriano. *Archaeological Guide to Rome.* Milan: Electa, 2005.

Lee, A. D. *Information and Frontiers: Roman Foreign Relations in Late Antiquity.* Cambridge: Cambridge University Press, 1993.

Lendon, J. E. *Soldiers and Ghosts: A History of Battle in Classical Antiquity.* New Haven, CT: Yale University Press, 2005.

Lewis, R.W.B. *Dante.* New York: Penguin, 2001.

Libanius. *Autobiography and Selected Letters.* Edited and translated by A. F. Norman. Cambridge, MA: Harvard University Press, 1992.

Light, Paul. *The True Size of Government.* Washington, DC: Brookings Institution, 1999.

Linowes, David F. *Privatization: Toward More Effective Government.* Urbana and Chicago: University of Illinois Press, 1988.

Lintott, Andrew. *Imperium Romanum: Politics and Administration.* London: Routledge, 1993.

Livy. *History of Rome.* Vol. 2. Translated by B. O. Foster. Loeb Classical Library. Cambridge, MA: Harvard University Press, 1916.

Llewellyn, Peter. *Rome in the Dark Ages.* New York and Washington, DC: Praeger, 1971.

Luttwak, Edward N. *The Grand Strategy of the Roman Empire: From the First*

Century A.D. *to the Third.* Baltimore: Johns Hopkins University Press, 1976.

Macdowall, Simon. *Adrianople,* A.D. *378.* London: Praeger, 2005.

MacMahon, A., and J. Price, eds. *Roman Working Lives and Urban Living.* Oxford: Oxbow Books, 2005.

MacMullen, Ramsay. *Corruption and the Decline of Rome.* New Haven, CT: Yale University Press, 1988.

———. *Romanization in the Time of Augustus.* New Haven, CT: Yale University Press, 2000.

———. *Roman Social Relations.* New Haven, CT: Yale University Press, 1974.

Madison, James [published anonymously]. *Letters of Helvidius: Written in Reply to Pacificus, on the President's Proclamation of Neutrality, Published Originally in the Year 1793.* Philadelphia: Samuel H. Smith, 1796.

Maier, Charles. *Among Empires: American Ascendancy and Its Predecessors.* Cambridge, MA: Harvard University Press, 2006.

Marcus Aurelius. *Meditations.* Translated by Maxwell Staniforth. London: Penguin, 1964.

Mathisen, Ralph W., and Hagith S. Sivan, eds. *Shifting Frontiers in Late Antiquity.* Aldershot, Hampshire: Variorum, 1996.

Mattern, Susan. *Rome and the Enemy: Imperial Strategy in the Principate.* Berkeley: University of California Press, 1999.

McDougall, Walter A. *Promised Land, Crusader State: The American Encounter with the World Since 1776.* New York and Boston: Houghton Mifflin, 1996.

McGregor, James H. *Rome from the Ground Up.* Cambridge, MA: Harvard University Press, 2005.

Meiggs, Russell. *Roman Ostia.* Second edition. London: Oxford University Press, 1973.

———. *Trees and Timber in the Ancient Mediterranean World.* Oxford: Clarendon Press, 1982.

Millar, Fergus. *The Emperor in the Roman World.* London: Duckworth, 1977.

National Commission on Terrorist Attacks upon the United States. New York: W. W. Norton, 2004.

Neustadt, Richard E., and Ernest R. May. *Thinking in Time: The Uses of History for Decision Makers.* New York: Free Press, 1986.

Nicolet, Claude. *Space, Geography, and Politics in the Early Roman Empire.* Ann Arbor: University of Michigan Press, 1991.

Niebuhr, Reinhold. *The Irony of American History*. New York: Scribner, 1952.

Nolan, Christopher, and David S. Goyer. *Batman Begins: The Screenplay*. London and New York: Faber and Faber, 2005.

Nordholt, Jan Willem Schulte. *The Myth of the West: America as the Last Empire*. Grand Rapids, MI: Eerdmans, 1995.

Norton, J. E. *The Letters of Edward Gibbon*. London: Cassell, 1956.

Nye, Joseph S., Jr. *The Paradox of American Power*. New York: Oxford University Press, 2002.

Origen. *Contra Celsum*. Translated by Henry Chadwick. Cambridge: Cambridge University Press, 1953.

Ormerod, Paul. *Why Most Things Fail: Evolution, Extinction & Economics*. New York: Pantheon, 2005.

Osborne, David, and Ted Gaebler. *Reinventing Government: How the Entrepreneurial State Is Transforming the Public Sector*. Reading, MA: Addison-Wesley, 1992.

Ovid. *Ovid in Six Volumes*. Vol. 6, second edition. Translated by Arthur Leslie Wheeler. Cambridge, MA: Harvard University Press, 1988.

Packenham, Robert A. *Liberal America and the Third World*. Princeton, NJ: Princeton University Press, 1973.

Packer, George. *The Assassins' Gate: America in Iraq*. New York: Farrar, Straus and Giroux, 2005.

Patten, Chris. *Cousins and Strangers*. New York: Times Books, 2006.

Perrottet, Tony. *Pagan Holiday: On the Trail of Ancient Roman Tourists*. New York: Random House, 2002.

Perry, William J., et al. *The U.S. Military: Under Strain and at Risk*. Washington, DC: National Security Advisory Group, 2006.

Pliny (the Younger). *Letters*. Vol. 2. Translated by Betty Radice. Loeb Classical Library. Cambridge, MA: Harvard University Press, 1969.

Pliny (the Elder). *Natural History*. Translated by H. Rackham. Loeb Classical Library. Cambridge, MA: Harvard University Press, 1940.

Plutarch. *Plutarch's Lives*. Vols. 2 and 3. Translated by Bernadotte Perrin. Loeb Classical Library. Cambridge, MA: Harvard University Press, 1914–1916.

———. *Plutarch's Lives*. Vol. 8. Translated by Bernadotte Perrin. Loeb Classical Library. New York: Macmillan, 1964.

Polybius. *The Histories*. Translated by W. R. Paton. Loeb Classical Library. Cambridge, MA: Harvard University Press, 1922.

Potter, David S. *The Roman Empire at Bay: AD 180–395*. New York: Routledge, 2004.

Ricci, David M. *The Transformation of American Politics: The New Washington and the Rise of Think Tanks.* New Haven, CT: Yale University Press, 1993.

Rickman, Geoffrey. *The Corn Supply of Ancient Rome.* Oxford: Clarendon Press, 1980.

Rodger, N.A.M. *The Command of the Ocean.* New York: W. W. Norton, 2005.

Sallust. *Sallust. The Jugurthine War and the Conspiracy of Cataline.* Translated by S. A. Handford. New York: Penguin, 1963.

Salway, Peter. *Roman Britain.* Oxford: Clarendon Press, 1981.

Schiavone, Aldo. *The End of the Past: Ancient Rome and the Modern West.* Translated by Margery J. Schneider. Cambridge, MA: Harvard University Press, 2000.

Schweitzer, Albert. *The Quest of the Historical Jesus.* First Complete Edition. London: SCM Press, 2000.

The Scriptores Historiae Augustae. Vol. 1. Translated by David Magie. Loeb Classical Library. Cambridge, MA: Harvard University Press, 1921.

Seeley, J. R. *The Expansion of England.* London: Macmillan, 1971.

Shaw, George Bernard. *Caesar and Cleopatra.* New York: Penguin, 1951.

Sherwin-White, A. N. *The Letters of Pliny.* Oxford: Clarendon Press, 1966.

Singer, P. W. *Corporate Warriors: The Rise of the Privatized Military Industry.* Ithaca, NY: Cornell University Press, 2003.

Sogno, Cristiana. *Q. Aurelius Symmachus: A Political Biography.* Ann Arbor: University of Michigan, 2006.

Steel, Ronald. *Walter Lippmann and the American Century.* Boston: Atlantic Monthly Press, 1980.

Sterling, Bruce. *Holy Fire.* New York: Bantam, 1996.

Suetonius. *Suetonius.* Translated by J. C. Rolfe. Loeb Classical Library. Cambridge, MA: Harvard University Press, 1970.

Swain, Joseph Ward. *Edward Gibbon the Historian.* New York: St. Martin's, 1966.

Sweig, Julia E. *Friendly Fire: Losing Friends and Making Enemies in the Anti-American Century.* New York: Public Affairs, 2006.

Tacitus. *Tacitus.* Vol. 1. Translated and revised by M. Hutton, R. M. Ogilvie, E. H. Warmington, Sir W. Peterson, and M. Winterbottom. Loeb Classical Library, revised edition. Cambridge, MA: Harvard University Press, 1970.

———. *Tacitus.* Vols. 3, 4, and 5. Translated by John Jackson. Loeb Classical Library. Cambridge, MA: Harvard University Press, 1931–1937.

————. *The Annals of Imperial Rome.* Translated and with an introduction by Michael Grant. New York: Penguin, 1996.

Thompson, E. A. *The Huns.* Revised and with an afterword by Peter Heather. Oxford: Blackwell, 1996.

Thompson, E. A., trans. *A Roman Reformer and Inventor: Being a New Text of the Treatise De Rebus Bellicis.* Chicago: Arles, 1996.

Thucydides. *History of the Peloponnesian War.* Translated by Rex Warner. Penguin Classics, revised edition. New York: Penguin, 1972.

Tillyard, E.M.W. *The Elizabethan World Picture.* New York: Vintage, 1959.

Tocqueville, Alexis de. *Democracy in America.* New York: Knopf, 2003.

Todd, Emmanuel. *After the Empire: The Breakdown of the American Order.* New York: Columbia, 2003.

Vance, William L. *America's Rome.* 2 vols. New Haven, CT: Yale University Press, 1989.

Vinz, Warren Lang. *Pulpit Politics: Faces of American Protestant Nationalism in the Twentieth Century.* Albany: State University of New York, 1997.

Von Hagen, Victor W. *The Roads That Led to Rome.* New York: World Publishing, 1967.

Walsh, Kenneth T. *Air Force One: A History of Presidents and Their Planes.* New York: Hyperion, 2003.

Ward-Perkins, Bryan. *The Fall of Rome and the End of Civilization.* Oxford: Oxford University Press, 2005.

Washburn, Jennifer. *University, Inc.: The Corporate Corruption of American Higher Education.* New York: Basic Books, 2005.

Webster, Graham. *The Roman Imperial Army.* New York: Funk & Wagnalls, 1969.

Wells, Peter S. *The Battle That Stopped Rome: Emperor Augustus, Arminius, and the Slaughter of the Legions in the Teutoburg Forest.* New York: W. W. Norton, 2003.

Whittaker, C. R. *Frontiers of the Roman Empire: A Social and Economic Study.* Baltimore, MD: Johns Hopkins University Press, 1994.

————. *Rome and Its Frontiers: The Dynamics of Empire.* London: Routledge, 2004.

Willcox, William B., ed. *The Papers of Benjamin Franklin.* Vol. 24. New York and London: Yale University Press, 1984.

Williams, Derek. *The Reach of Rome.* New York: St. Martin's, 1996.

————. *Romans and Barbarians: Four Views from the Empire's Edge.* New York: St. Martin's, 1999.

Williams, Stephen. *Diocletian and the Roman Recovery.* New York: Routledge, 1997.

Wills, Garry. *Cincinnatus: George Washington & the Enlightenment.* New York: Doubleday, 1984.

Zwick, Jim. *Mark Twain's Weapons of Satire: Anti-Imperialist Writings on the Philippine-American War.* Syracuse, NY: Syracuse University Press, 1992.